K. E. Kemper · W. Blomquist · A. Dinar (Eds.)

Integrated River Basin Management through Decentralization

Karin E. Kemper · William Blomquist · Ariel Dinar (Eds.)

Integrated River Basin Management through Decentralization

With 18 Figures and 35 Tables

 Springer

Editors

Karin E. Kemper

The World Bank
1818 H St, NW
Washington, DC 20433, USA
E-mail: kkemper@worldbank.org

Ariel Dinar

The World Bank
1818 H St, NW
Washington, DC 20433, USA
E-mail: adinar@worldbank.org

William Blomquist

Department of Political Science
Indiana University-Purdue University Indianapolis (IUPUI)
425 University Blvd.
Indianapolis, IN 46202, USA
E-mail: blomquis@iupui.edu

Technical Editor

John Dawson

PO Box 15243
Langata 00509
Nairobi, Kenya
E-mail: jdawson@iconnect.co.ke

Library of Congress Control Number: 2006936101

ISBN-10   3-540-28354-4   Springer Berlin Heidelberg New York
ISBN-13   978-3-540-28354-6   Springer Berlin Heidelberg New York

Springer is a part of Springer Science+Business Media
springer.com
© Springer-Verlag Berlin Heidelberg 2007

Cover design: E. Kirchner, Heidelberg
Typesetting: Uwe Imbrock, Stasch · Verlagsservice, Bayreuth (*stasch@stasch.com*)
Production: Agata Oelschläger

Printed on acid-free paper   30/2132/AO – 5 4 3 2 1 0

Foreword

This volume deals with three critical development and poverty reduction issues: water, the linkages between sustainable natural resources management and infrastructure, and participatory decisionmaking. Since the beginning of time, human beings have tried to harness and manage water for multiple uses, ranging from irrigation and shipping to drinking water supply and hydropower. In the past century it has become clear that global population growth and economic development put increasing stress on the water resources base. By 2035 as many as three billion people, almost all of them in developing countries, could live under conditions of severe water stress, especially if they happen to live in Africa, the Middle East, and South Asia. This will cause obvious hardship but it will also hold back the economic growth needed for millions of people to escape poverty.

To sustain this vital resource, we need to find technical, institutional, social and economically viable solutions.

This pioneering work comprehensively analyzes development outcomes of decentralization of decisionmaking from central to local levels in water resources management. The book includes a global analysis of 83 river basin management organizations, complemented by case studies spanning from Latin America to Asia and Australia.

The results are both encouraging, and challenging. On the one hand there is a clear recognition that institutional change is slow and that the political economy in the water sector plays a key role in the potential of achieving sustainable outcomes. On the other hand the results are also encouraging, because the analysis shows the power of humankind's ingenuity in addressing the problems we face, be they water scarcity, drought, floods, or pollution. The case examples analyzed in this book show that when there is leadership and vision, institutions can be adapted to different local conditions, and as a result complex issues can be addressed and solved.

I am pleased that this book comes at this particular moment, when the global community more than ever realizes the need for integrated and more comprehensive approaches to development through the involvement of concerned groups in decisionmaking. I hope that it will serve decisionmakers, practitioners and those interested in water resources management in providing inspiration for the development of their specific institutional needs.

Katherine Sierra
Vice President, Sustainable Development, The World Bank

Acknowledgments

This book is a product of the study Integrated River Basin Management and the Principle of Managing Water Resources at the Lowest Appropriate Level – When and Why Does It (Not) Work in Practice? The study was led by Karin Kemper and Ariel Dinar from the World Bank and William Blomquist from Indiana University. The editors would like to acknowledge the many individuals and organizations that were involved in making this work possible as it evolved over the time period 2002 to 2006.

Special thanks are due to our team members Michele Diez, Rosa Maria Formiga Johnsson, William Fru, Gisèle Sine, Corazon Solomon, Oliver Taft, Catherine Tovey, Sadaf Alam, and Melissa Williams, who supported various components of this project. The International Network of Basin Organizations was instrumental in helping us identify and get in touch with the more than 120 river basin organizations that constitute the basis for the global analysis.

Certainly, this book would not have become as rich without the excellent contribution of the basin case study and background paper authors who, most of the time, also doubled as organizers – and sometimes interpreters – during our local visits: Maureen Ballestero (Tárcoles – Costa Rica), Ken Calbick and David Marshall (Fraser – Canada), Rosa Maria Formiga Johnsson (Alto Tietê and Jaguaribe – Brazil), Consuelo Giansante (Guadalquivir – Spain), Brian Haisman (Murray Darling – Australia), Kikkeri Ramu and Trie Mulat Sunaryo (Brantas – Indonesia), Andrzej Tonderski (Warta – Poland), and Jyothsna Mody (Accountability through Decentralization).

We would like to highlight the wonderful technical editing done by John Dawson. He managed to make a homogeneous product out of the many different background papers and chapter drafts, some of which had authors from as many as three different language backgrounds and continents. Thanks are due also to Uwe Imbrock, who made our interactions with Springer as smooth and efficient as possible.

We are immensely grateful to the individuals whom we interviewed in the course of this research, to the many colleagues who reviewed and commented on earlier versions of the various papers, and to the participants in the workshop in Warsaw, Poland, in May 2005, who provided such important insights and feedback. None of those individuals is responsible for the findings and conclusions in this book. The views expressed in this document are those of the authors and should not be attributed to the World Bank.

The study was funded by the World Bank's Research Committee, the Agriculture and Rural Development Department, the Water Resources Management Group, and the

South Asia Social and Environment Unit. We also appreciate the financial support of the Bank Netherlands Water Partnership Program (BNWPP).

Further background documentation for this book can be found at *www.worldbank.org/riverbasinmanagement*.

Karin Kemper, William Blomquist, and Ariel Dinar
November 2006

Contents

Contributors

Maureen Ballestero
Apartado Postal 14-5000
Liberia, Guanacaste, Costa Rica
Email: tempis@racsa.co.cr

Maureen Ballestero received a degree in Agricultural Economics from the University of Costa Rica and a Master's degree in Business Administration from the Latin University of Costa Rica. Since 1999 she has been the Regional Coordinator of the Central American Advisory Committee of the Global Water Partnership (GWP-CATAC). She is also the director of the Management Project of the Tempisque River Basin, located in the province of Guanacaste, Costa Rica. Maureen is also a private consultant on topics such as integrated water resources management, water resources planning, policies, and institutional and legal frameworks. Currently she is working as a consultant for the Inter-American Development Bank (IADB), coordinating the National Integrated Water Resources Management Strategy in Costa Rica.

Anjali Bhat
Junior Research Fellow
Natural Resources and Social Dynamics
ZEF (Centre for Development Research)
Walter Flex Strasse 3
53113 Bonn, Germany
E-mail: abhat@uni-bonn.de

Anjali Bhat has a B.S. in Environmental Science from the University of Illinois, Champaign-Urbana, and an MPA from Indiana University, where she was affiliated with the Workshop in Political Theory and Policy Analysis. Her research interests are centered around environmental governance and the decentralization of authority in natural resources management. She has consulted for the World Bank, the International Food Policy Research Institute, and the World Resources Institute, and is currently a Junior Research Fellow at the Center for Development Research at the University of Bonn. Her doctoral research, supported by a Fulbright scholarship, examines the impact of decentralization policies upon integrated river basin management in Indonesia.

William Blomquist
Department of Political Science
Indiana University-Purdue University Indianapolis (IUPUI)
425 University Blvd., Room 503L
Indianapolis, IN 46202, USA
E-mail: blomquis@iupui.edu

William Blomquist is Professor of Political Science at Indiana University Purdue University Indianapolis (IUPUI). His research interests concern institutions and policies for managing water resources. He is the author or co-author of several research publications on these topics, including the books Dividing the Waters (1992) and Common Waters, Diverging Streams (2004). He received his B.Sc. in Economics and M.A. in Political Science from Ohio University, and his Ph.D. in Political Science from Indiana University.

Ken S. Calbick
20828 McFarlane Avenue
Maple Ridge, BC
Canada, V2X 7S1
E-mail: kcalbick@shaw.ca

Ken Calbick earned his Bachelors of Technology Honors degree in environmental engineering from the British Columbia Institute of Technology. He has earned a Master's Degree in resource management from Simon Fraser University, and is currently pursuing a Ph.D. in the School of Resource and Environmental Management for which he has received a prestigious Canada Graduate Doctoral Scholarship from the Social Sciences and Humanities Research Council of Canada. Additionally, he has 14 years experience within the environmental field encompassing air quality issues, liquid waste management, solid waste handling, hazardous waste disposal, waste reduction and recycling initiatives, and conflict assessment and resolution. Ken also has eight years of project management experience in a variety of settings, including the public sector, private industry, consulting, and academia.

John Dawson (Technical Editor)
PO Box 15243
Langata 00509
Nairobi, Kenya
E-mail: jdawson@iconnect.co.ke

John Dawson is a freelance writer and editor based in Nairobi, Kenya. A native of Liverpool, England, and a graduate of Oxford University, he taught in Oxford, Cairo and Nairobi before finally succumbing to the enticing appeals of the English language. His two published books, Quiz Setting Made Easy and East Africa Alive, reflect his interest in mind puzzles and wildlife. He edits, upgrades, writes and rewrites documents for a number of international organizations, the World Bank included, and attends conferences as a report writer for the United Nations. During his varied working life he has been editor of three magazines and technical editor of several books, has written numerous published articles, and has worked as a professional photographer. He plays whistle and accordion for an Irish group in Nairobi.

Ariel Dinar
The World Bank
1818 H St, NW
Washington, DC 20433, USA
E-mail: adinar@worldbank.org

Ariel Dinar is a Lead Economist with the Agriculture and Rural Development Department of the World Bank. He earned his Ph.D. from the Hebrew University of Jerusalem in Israel in 1985 in Agricultural and Resource Economics. He works on, publishes and teaches various aspects of water economics and institutions of water resources at local, regional and international levels. His recent co-authored books include The Institutional Economics of Water: A Cross Country Analysis of Institutions and Performance (2004), and Bridges over Water: Understanding Transboundary Water Conflict, Negotiation and Cooperation (2007).

Rosa Maria Formiga Johnsson
Rua Nascimento Silva, 91 apto 102, Ipanema
Rio de Janeiro-RJ Brazil
CEP: 22.421-020
Email: aformiga@terra.com.br

Rosa Maria Formiga Johnsson received her M.S. and Ph.D. in Environmental Sciences and Technologies from Université de Paris XII, France. She is currently an associate professor at the State University of Rio de Janeiro in the Department of Environmental and Sanitary Engineering. Between 1999 and 2006, she was an Associate Researcher at the Hydrology and Environmental Studies Laboratory of the Federal University of Rio de Janeiro where she conducted research and policy analysis related to the implementation of water policy and management reforms in Brazil. She has been the executive secretary of the Watermark Project, a multidisciplinary, collaborative study of new water management institutions throughout Brazil (20 river basins) since 2000. She is also presently involved in research on bulk water pricing and integrated water resource management.

Consuelo Giansante
c/ San Blas11
41003 Seville, Spain
Email: giansante@terra.es

Consuelo Giansante is a consultant in water and natural resources management. Her background is in biology (B.Sc. Biology, University of Milan, Italy; DEA, Université Paris VII, France) and natural resources management (M.Sc. Conservation, University College London, United Kingdom). She has been research assistant at University College London and the University of Seville. She has been involved in several projects on integrated river basin management and drought management practices. She is currently an advisor to the Head of the Andalusian Water Agency.

Brian Haisman
18 Carramar Crescent
Winmalee NSW 2777, Australia
E-mail: brianhev@optusnet.com.au

Brian Haisman is a consulting water resources manager with an independent practice specializing in all aspects of applied water policy and integrated river basin management, particularly water rights, institutional analysis, and asset management strategies. He holds a Master of Engineering Science from the University of New South Wales, Australia. He has lengthy experience in community-based watershed management, has held senior positions in the Murray-Darling Basin Commission and was Director, Water Resource Management for the lead New South Wales water agency. He has been awarded the Order of Australia Public Service Medal for outstanding services to water conservation and natural resource management.

Karin E. Kemper
The World Bank
1818 H St, NW
Washington, DC 20433, USA
E-mail: kkemper@worldbank.org

Karin E. Kemper is a Lead Water Resources Management Specialist at the World Bank. She has extensive work experience in the water sector and has carried out project investment activities, studies and research in a variety of countries. Her special interests include institutional and economic aspects of water management, including river basin and groundwater management, water allocation mechanisms and instruments, transboundary water issues, water user participation and decentralization. A native of Germany, she holds a Ph.D. in Water and Environmental Studies from Linköping University in Sweden.

Kikkeri Ramu
7811 Heritage Farm Dr
Montgomery Village
Maryland 20886, USA
Email: kvramu32@yahoo.com or kvramu@aol.com

Kikkeri Ramu has worked with the World Bank as a consultant since 1998 on the Bank's water sector programs in Indonesia. He specializes in river basin water resources management and institutional frameworks. Prior to this, he was an advisor on water sector policies and programs to the Directorate General of Water Resources, Indonesia, from 1990. From 1975 to 1989, as Team Leader/Project Director for ECI/USA, he was involved in the implementation of a number of irrigation and water resources projects in Indonesia. Between 1954 and 1971 he worked as a field engineer on water projects and as a faculty member in India. Dr Ramu has a Ph.D from Colorado State University, USA, and a BS (Hons) and MS from India.

Andrzej Tonderski
Executive Director
POMCERT-Pomeranian Center for Environmental Research and Technology
Sobieskiego 18/19, PL 80-952
Gdansk, Poland
E-mail: pomcert@univ.gda.pl

Andrzej Tonderski is a graduate of Gdansk University of Technology (Environmental Engineering). His Ph.D. at Linköping University (Department of Water and Environmental Studies) in Sweden was about the Control of Nutrient Fluxes in Large River Basins, especially in the Baltic Sea Drainage Basin. For many years he has been working as a consultant (AF International, ENVISTON, SMHI, Sweden) on international water management issues. Since 2003 he has been responsible for development and management of the Pomeranian Center for Environmental Research and Technology - POMCERT) in Gdansk, Poland.

Acronyms and Abbreviations

List of acronyms and abbreviations

ACOPE	Costa Rican Association of Energy Producers (Asociación Costarricense de Productores de Energía)
ANA	National Water Agency (Agência Nacional de Águas)
ASOCUENCAS	Association for Hydrographic River Basins (Asociación para Cuencas Hidrográficas)
ASSEMAE	National Association of Municipal Sanitation Services (Associação Nacional dos Serviços Municipais de Saneamento)
AyA	Institute of Aqueducts and Sewers (Instituto Costarricense de Acueductos y Alcantarillados)
Balai PSDA	basin water resources management unit (balai pengelolaan sumber daya air)
CA	comunidad autónoma
CAGECE	Ceará State Water and Sanitation Company (Companhia de Água e Esgoto do Estado do Ceará)
CEDERENA	Center for Environmental Law and Natural Resources (Centro de Derecho Ambiental y de Los Recursos Naturales)
CETESB	São Paulo State Environment Agency (Companhia de Tecnologia de Saneamento Ambiental)
CH	confederación hidrográfica
CNFL	National Power and Light Company (Compañía Nacional de Fuerza y Luz)
COAG	Council of Australian Governments
COGERH	Water Resource Management Company (of Ceará) (Companhia de Gestão dos Recursos Hídricos)
CORHI	Coordinating Committee for the State Water Resources Plan (Comitê Coordenador do Plano Estadual de Recursos Hídricos)
CRGT	Commission for the Río Grande de Tárcoles Basin (Comisión de la Cuenca del Río Grande de Tárcoles)
CRH	State Water Resources Council (Conselho Estadual de Recursos Hídricos)
DAEE	Department of Water and Electric Energy (Departamento de Água e Energia Elétrica)

Continued

DDWM	district directorate of water management
Dinas PUP	Provincial Water Resource Services Office
DNOCS	National Department of Drought Relief (Departamento Nacional de Obras contra as Secas)
DOU	User Mobilization Department (Departamento da Organização de Usuários)
EMAE	Metropolitan Water and Energy Company (Empresa Metropolitana de Águas e Energia)
EMASESA	Empresa Municipal de Aguas de Sevilla, SA
EMPLASA	Metropolitan Planning Agency for Greater São Paulo (Empresa Metropolitana de Planejamento da Grande São Paulo)
FBMB	Fraser Basin Management Board
FBMP	Fraser Basin Management Program
FECON	Federation of Environmental Groups (Federación Costarricense de Conservación del Ambiente)
FEHIDRO	State Fund for Water Resources (Fundo Estadual de Recursos Hídricos)
Feragua	Guadalquivir Irrigation Farmers Union (Federación de Regantes del Guadalquivir)
FNS	Federal Health Agency (Fundação Nacional de Saúde)
FREMP	Fraser River Estuary Management Program
FUDEU	Foundation for Urban Development (Fundación para el Desarrollo Urbano)
FUNCEME	Foundation for Meteorology and Water Resources of Ceará (Fundação Cearense de Metereologia e Recursos Hídricos)
ICE	Costa Rican Electricity Institute (Instituto Costarricense de Electricidad)
IDB	Inter-American Development Bank
IMF	International Monetary Fund
INBO	International Network of Basin Organizations
JICA	Japan International Cooperation Agency
MINAE	Ministry of Environment and Energy (Ministerio del Ambiente y Energía)
NBWM	National Board of Water Management
NCWM	National Council of Water Management
PC	principal component (analysis)
PDAM	domestic water supply company
PJP II	Pembangunan Jangka Panjang (Indonesia 25-year development plan)
PJT I	Brantas River Basin Management Corporation (Perum Jasa Tirta I)
PLN	National Power Corporation

Continued

PPTPA	Basin Water Resources Committee
PTPA	Provincial Water Resources Committee
RBWM	regional board of water management
RCWM	regional council of water management
RWMA	regional water management authority
SABESP	Basic Sanitation Company of São Paulo State (Companhia de Saneamento Básico do Estado de São Paulo)
SEMACE	Ceará State Environment Superintendancy (Superintendência Estadual do Meio Ambiente do Ceará)
SERH	Energy, Water Resources, and Sanitation Secretariat (Secretaria de Energia, Recursos Hídricos e Saneamento)
SINAC	National System of Conservation Areas (Sistema Nacional de Áreas de Conservación)
SMA	State Secretariat for the Environment (Secretaria de Estado do Meio Ambiente)
SOHIDRA	Hydraulic Works Superintendency (Superintendência de Obras Hidráulicas do Estado do Ceará)
SRH	Secretariat for Water Resources (Secretaria dos Recursos Hídricos)
WUA	water user association
WUAF	federation of water user associations

Part I

**River Basin Management
at the Lowest Appropriate Level:
When and Why Does It Work
in Practice?**

Chapter 1

River Basin Management at the Lowest Appropriate Level: When and Why Does It (Not) Work in Practice?

K. E. Kemper · W. Blomquist · A. Dinar

1.1
Motivation for This Study

When the research project that has led to the material and analysis presented in this book started, a great deal of investigation had been carried out into the application of the four so-called Dublin Principles of 1992 (ICWE 1992). These are frequently quoted in the water literature and have been guiding much of the thinking about water resource management in the past one and a half decades. Most discussion, however, has taken place around three of the Dublin Principles: those related to water as an economic good, the role of women in provision and management of water, and the need for integrated water resource management.

Interestingly, the fourth principle, which concerns river basin management at the lowest appropriate level, was also being promoted and applied, but it was more or less taken for granted that it was a desirable practice, with little enquiry into whether it really worked and what the outcomes of its application were. These questions are, of course, vital for policymakers and water users throughout the world, especially in light of the number of river basin management efforts that are under way in the 21st century. Governments in several countries, multilateral financing agencies such as the World Bank, and other institutions such as the Global Water Partnership promote river basin organizations as a means of advancing river basin management at the lowest appropriate level. Accordingly, a study was carried out to consider those questions in a systematic way; this book presents the outcomes of the investigations into this issue.

1.2
Study Approach

In order to investigate whether river basin management at the lowest appropriate level really works and what the outcomes are when it is applied, a three-tiered approach was developed. First, a literature review was carried out (Mody 2004), both to identify cases where the issue had already been studied, and to better define what the principle means. Based on that literature review, the study team developed an analytical framework intended to capture the factors likely to be related to river basin management success and generate hypotheses that could be tested in actual settings where river basin management had been attempted. That analytical framework is presented later in this chapter.

Second, with the help of the International Network of Basin Organizations (INBO) as many river basin organizations as possible were identified (a total of 197 worldwide) and a survey was constructed, based on the analytical framework, that would permit the history, objectives, process, and performance of water resource management at the basin level to be traced in a structured manner. For the statistical analysis 83 surveys were retained, based on the quality of data received (see Table 3.1).

Third, in order to complement the quantitative findings of the survey and explore questions that are difficult to answer through survey research, a qualitative case study was designed. Eight river basins were selected from across the world that exhibited different characteristics in terms of economic development, water resource management needs and challenges, success in river basin management, basin management structures, and processes to arrive at these structures. These are basins where organizations have been developed at the basin or subbasin scales to perform (or coordinate the performance of) management functions such as planning, allocation and pricing of water supplies, flood prevention and response, and water quality monitoring and improvement. Members of the study team visited these river basins (the Murray-Darling basin in Australia, the Jaguaribe and Alto Tietê basins in Brazil, the Fraser basin in Canada, the Tárcoles basin in Costa Rica, the Brantas basin in Indonesia, the Warta basin in Poland, and the Guadalquivir basin in Spain) over the time period 2003 to 2005 and carried out in-depth discussions for the purpose of this study. The cases have been compared and assessed for their observed degrees of success in achieving improved stakeholder participation and integrated water resource management.

1.3
What Does "River Basin Management at the Lowest Appropriate Level" Mean, and Why Is It Important?

In the majority of countries, river basin management has traditionally been the mandate of government entities, such as federal or national water resource agencies or ministries. By the end of the 1980s, it was clear that this approach often did not work well and did not produce the desired results, especially in developing countries.

Analyses pointed to a need for decentralization of decisionmaking and the active involvement of stakeholders, the assumption being that decisions taken by and with stakeholders would be better informed and would allow negotiation among stakeholder groups in order to come to more rational and equitable solutions. Such processes might also lower resistance to sometimes difficult decisions. An important point in this regard is the phrase "lowest appropriate level", implying that there is no prescription with regard to full decentralization at all costs, but the principle implies locally adapted decentralization of decisionmaking, with some decisions being devolved to stakeholders, and others being kept at central, state, or provincial levels, when and as appropriate.

Perhaps somewhat underestimated, decentralization – and especially decentralization in water resource management – is quite a radical concept. Water allocation is a political issue in most countries, especially the more arid ones, and is frequently dominated by vested interests. Decentralization therefore is a reform process that changes entitlements and may lead to winners and losers. It may imply more security for some (notably those who before the decentralization had no information about their water resource and no voice in allocation issues) and less security for others (specifically those who could always count on getting their allocation from the central authority, either by right, by payoffs, or by influence). Thus, while decentralization of decisionmaking overall can improve performance of water resource management (as will be shown in the following chapters), there clearly are social and political tradeoffs that need to be taken into account in its implementation.

For this reason, the question when and why decentralization functions in practice, or why it does not, is of major practical relevance. For most countries, functioning water resource management is an essential basis for growth. This relates to the institutional structures for planning and operation of existing infrastructure, and for financing these processes, as well as to the need for appropriate physical infrastructure to be constructed in a timely manner (Grey and Sadoff 2006). With the centralized approaches not functioning (for example due to too much infrastructure, investments in the wrong place or of wrong size, inadequate long-term financing and maintenance, negative externalities, or inequities in accrual of benefits from water resource management decisions),[1] the need for decentralization not only affects the water resource sector, but effectively the basis for growth of the countries in question. Case studies in this book, for instance, illustrate how decisionmaking in the Murray-Darling basin in Australia over time has allowed its residents to prosper (and is now helping them to deal with environmental issues that used to be neglected), and how traditionally ignored water users in the Jaguaribe basin in northeast Brazil now have more livelihood security due to the "negotiated allocation" system that has been instituted over the past decade. The studies also show, however, that without formalized decentralization the stakeholders in Costa Rica's Tárcoles basin have not been able to tackle the major pollution issues that undermine the environmental health of the country's economic heartland around the capital San José.

Consequently, the investigation into the outcomes of decentralization processes in water resource management, and into the factors that enable them to function, is more than an academic exercise. It is of fundamental importance to inform public debate and policy decisionmaking with regard to a crucial productive sector.

The following sections will describe the analytical framework used for the present study, followed by the analysis of the overall study results (combining the global survey and the case studies), and the policy recommendations.

1.4
Analytical Framework

An analytical framework has been developed to analyze the data gathered for the study. It identifies a number of political and institutional factors that may be associated with the emergence, sustainability, and success or failure of decentralized approaches to

---

[1] See for instance *Troubled Waters*, World Bank (2005), which provides an in-depth analysis of the highly centralized Indian water sector.

integrated water resource management at the basin scale. These factors, and their
hypothesized relationships with basin management in a country that has decentralized or is attempting to decentralize water resource management institutions,
are derived from the institutional analysis literature relating to water or other
natural resource management and to decentralized systems (especially Ostrom
1990, 1992; also Agrawal 2002; Alaerts 1999; Bromley 1989; Easter and Hearne 1993;
Wunsch 1991).

The institutional analysis framework was adopted because its focus on institutional
development as a collective outcome of individual choices made within a social,
cultural, and political-economic framework seemed best suited to a study intended
to illuminate the origin and evolution, as well as the performance, of river basin
institutions.

Four sets of variables were identified under the following major headings:

- Contextual factors and initial conditions
- Characteristics of the decentralization process
- Central-local relationships and capacities
- Basin-level institutional arrangements

These sets of variables are not directly linked to basin management success or
failure, but influence characteristics that are also believed to be linked to those
outcomes (Fig. 1.1).

Before the factors that influence performance are discussed, it is necessary
to define what would be considered improved performance of river basin management
itself. While the overall, desired outcome of improved institutional arrangements would
be sustainable and equitable water resource management in a river basin, each basin
has a different starting point, and generally each river basin organization has set
different objectives for itself. These can include:

- Improved water allocation across water uses
- Improved water quality
- Reduction of disputes over water allocation or water quality
- Reduction in loss of production or productivity due to water scarcity or flooding
- Increase in the basin's gross domestic product.

Accordingly, in the assessments of decentralization performance, whether the
decentralization process had led to the achievement of these objectives was measured
against the stated objectives of each river basin organization.

1.4.1
Contextual Factors and Initial Conditions

The literature on natural resource management indicates that successful decentralization
is at least partly a function of the initial conditions that prevail at the start of, and give
context to, a decentralization initiative. These include:

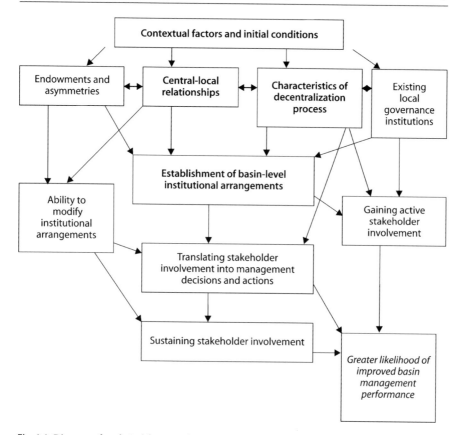

Fig. 1.1. Diagram of analytical framework (Note: *bold* = categories of variables detailed further below)

- Level of economic development in the nation
- Level of economic development of the river basin
- Distribution of resources among basin stakeholders
- Social and cultural distinctions among stakeholders.

Level of Economic Development in the Nation

Although decentralization may be undertaken in the hope of reducing the financial outlays of central government, it may, in the early stages of the transition, need to financially assist basin stakeholders in establishing organizations and practices that will facilitate basin management.

Also, aspects of water resource management that have the characteristics of public goods may continue to be efficiently provided by a central government, for example weather monitoring and forecasting, and perhaps also hydrological research, research on the health and environmental effects of water quality, and some aspects of flood control.

All other things being equal, decentralization initiatives are expected to be more likely to achieve sustainable success where the economic well-being of the nation allows the central government to bear transition costs and ongoing costs of functions that support and facilitate basin-scale management.

Level of Economic Development of the River Basin

Developing and maintaining decentralized basin-level institutional arrangements for water allocation, water quality protection, monitoring, and enforcement will require some commitment of financial and other resources from basin stakeholders, even where the central government provides transitional or ongoing financial support.

Furthermore, the literature on decentralized water resource management indicates that effective decentralization must include some degree of financial autonomy. Sustaining this autonomy often depends upon implementing some form of water pricing or tariffs, which entails requiring at least some water users to pay for a resource they previously consumed for free or at greater subsidization. Basins where stakeholders can sustain such commitments are (all other things being equal) more likely to achieve sustainable success.

Distribution of Resources among Basin Stakeholders

The initial distribution of resource endowments among the basin stakeholders is an important contextual factor in the development and successful implementation of a decentralization initiative, though its manifestations may be complex.

On the one hand extreme asymmetries in resource endowments can imperil decentralization success. Some stakeholders may have such financial and political control over the rights to basin resources that they might attempt to derail a decentralization process that leaves them less well off. Other stakeholders may be so destitute that they rationally elect not to participate in a potentially advantageous process of improved resource management if they are unable to bring any resources of their own.

On the other hand, some inequality of initial resource endowments may facilitate movement towards a successful and stable management regime by enabling some stakeholders to bear the costs of taking a leadership role if they see themselves having a substantial financial or managerial stake in the future of the resource (Blomquist 1988; Ostrom 1990).

Social and Cultural Distinctions among Stakeholders

Class, religious, or other social and cultural distinctions can affect successful implementation of decentralization initiatives through their effects on stakeholder communication, trust, and extent of prior experience in cooperative endeavors.

1.4.2
Characteristics of the Decentralization Process

Two necessary conditions of a successful decentralization initiative are devolution of authority and responsibility from the center, and acceptance of that authority and responsibility by local entities in the basin. Whether both occur will depend in part upon why and how the decentralization takes place. They can be captured in the following variables:

- Top-down, bottom-up, or mutually desired devolution
- Incorporation or involvement of existing local governance arrangements
- Consistent central government policy commitment

Top-Down, Bottom-Up, or Mutually Desired Devolution

Successful implementation of a decentralization initiative may depend significantly upon its motivation. In some cases, central government officials may have undertaken decentralization initiatives for ulterior motives – for example to reduce their own accountability, or to solve a budgetary crisis by reducing their financial responsibility in some areas, or as a precondition for continued external support. In other cases the decision to decentralize may have been the outcome of a process of mutual discussion and agreement between central officials hoping to improve performance outcomes and local stakeholders desiring greater autonomy and flexibility in managing the resource. Whatever the reality, such initiatives are more likely to be implemented successfully when devolution is desired mutually by basin stakeholders and central government officials.

Incorporation or Involvement of Existing Local Governance Arrangements

The literature suggests that stakeholder involvement, and the commitment of human and financial resources, is likely to be greater if traditional community governance institutions (for example at village and tribe level) and practices are recognized and incorporated into the devolution process. Also, the transaction costs (primarily in terms of time and effort) to basin stakeholders of relating to existing organizational forms are expected to be smaller than the costs of decentralization initiatives that feature central government construction of new basin-level organizations that are largely separate from existing community governance institutions. It is still the case, however, that some new organizations will have to be created in order to achieve basin-scale management through promotion of communication and the integration of decisionmaking across communities within a river basin.

Consistent Central Government Policy Commitment

Adoption and announcement of a decentralization policy – in the form of a statute or regulation, for instance – might occur swiftly in some cases, but in nearly all cases decentralization (which may entail adjustments to agency responsibilities, budgetary modifications, and so on) will take time, and may have to survive such potentially unsettling factors as changes in central government, which may or may not continue the existing policy regime. In the latter instance discontinuities in central government policy commitments can disrupt support, confuse the missions and operations of central government agencies involved in resource management, and undermine the confidence of stakeholders in the decentralization initiative.

1.4.3
Central-Local Relationships and Capacities

Because successful decentralization requires complementary central and local actions, other aspects of the central-local relationship and their respective capacities can be expected to condition that success. These include:

- The extent of actual devolution
- Financial resources and autonomy at the basin level
- Basin-level authority to create and modify institutional arrangements
- Local experience with self-governance and service provision
- Distribution of national-level political influence among stakeholders
- Characteristics of the water rights system
- Adequate time for implementation and adaptation.

The Extent of Actual Devolution

Announcement of a decentralization policy may in fact be only symbolic, with central government in practice retaining control over all significant resource management decisions; or it may represent an abandonment of central government responsibility for resource management without a concomitant establishment of local-level authority.[2] In such instances stakeholder willingness to commit to and sustain the active involvement necessary for success will be undermined (for example Vermillion and Garces-Restrepo 1998 on Colombia). The success of the devolution process is more likely in situations where stakeholders acquire both authority and responsibility for aspects of resource management.

Financial Resources and Autonomy at the Basin Level

Although decentralization of resource management means at least some assumption of financial responsibilities by basin organizations and stakeholders, it does not have to mean that they become solely responsible for all resource management funding. One of the indicators of central government support for decentralization can be its

---

[2] This is distinct from the discussion in the previous section about the motivation for the decentralization policy. A central government might adopt a decentralization policy out of sincere motivations and in consultation with local stakeholders, but nevertheless fail for other reasons to actually relinquish any control over resource management or establish local authority.

willingness to provide some financial assistance to basin-level organizations without intrusive control over basin-level decisions about the priorities on which those funds shall be spent. It is expected that the greatest prospects for success lie somewhere toward the middle of a spectrum where complete central government funding and control lie at one pole and complete basin-level funding and control lie at the other.[3]

Basin-Level Authority to Create and Modify Institutional Arrangements

Institutional arrangements at the basin level – for governance, financing, monitoring, infrastructure construction and maintenance, and so on – are more likely to function effectively if tailored to the particular physical, social, and economic setting of each basin. They are also more likely to function effectively for long periods – to be sustainable – if they can be modified in response to changed conditions. Thus a key element is the extent to which local communities can design and implement their own institutional arrangements, for two reasons: first, they are more likely than central government officials to be able to deal with the information requirements of such tasks; and second, the ability to craft their own institutional arrangements can be expected to attract more active involvement from basin-level stakeholders.[4]

Local Experience with Self-Governance and Service Provision

Because there is no established recipe for organizing water resource management at the lowest appropriate level, the ability of central government officials to strike a balance between supportiveness and intrusiveness, and the capacity of stakeholders to organize and sustain institutional arrangements, will depend in part on their experiences with other public services or responsibilities.

---

[3] An optional further refinement of this variable would involve inquiring not only about the gross proportions of funding between central and local agencies but about how funding relates to resource management functions. In this refinement, successful implementation of decentralization in water resource management would be linked to central government funding and control of functions that are best organized at a larger scale because of their high capital or technical requirements (for example, conducting hydrogeological surveys or developing water quality standards) and basin-level funding and control of functions that are best organized on a smaller scale because of their time-and-place specificity or sociocultural implications (for example, decisions about allocation of water between different sectors, monitoring of individuals' or households' water use, maintenance of infrastructure facilities).

[4] Two extensions or implications of this logic should be mentioned. First, the creation of effective basin-scale arrangements will often require the ability to establish cross-jurisdictional or inter-jurisdictional institutions, since basin boundaries may not conform neatly to the boundaries of existing political or administrative jurisdictions. Accordingly, the central-local relationship confers greater autonomy on basin-level stakeholders to create and modify institutional arrangements for decentralized water resource management if it includes the authority to develop cross-jurisdictional arrangements than if it does not. Second, the authority to create and modify institutional arrangements may not extend only to basin-scale arrangements. Basin-level stakeholders may also discern the need for or utility of some subbasin arrangements, and the central-local relationship confers greater autonomy on basin-level stakeholders if it also allows them to create and modify these as well.

One limitation of this logic should also be mentioned. The more organizations stakeholders create, the greater the transaction costs of maintaining all of them and coordinating their activities. Stakeholders may therefore quite rationally opt to rely on some existing institutions to conduct management functions, even though boundary and other considerations seem to be less than perfectly fitted to the task, in order to limit transaction costs to a manageable level.

We would expect that water resource management decentralization initiatives are more likely to be implemented successfully in settings where local participants have experience in governing and managing other resources or public services – for example land uses, schooling, and transportation – and are already practiced at raising, maintaining, and distributing revenues, resolving disagreements, and taking collective decisions.

Distribution of National-Level Political Influence among Stakeholders

The effectiveness of basin-level institutional arrangements will be a function of stakeholder commitment to them and willingness to abide by decisions or actions taken by basin-level institutions. That commitment can be undermined if some stakeholders can exert undue influence with central officials to exempt them from certain basin management decisions, or structure advantages on their behalf; then bargaining among stakeholders over resource management decisions will not be conducted in good faith and resource management policies will not be enforced fairly. All other things being equal, it is expected that successful implementation of decentralized management would more often be found where stakeholders have relatively symmetrical political influence with central government.

Characteristics of the Water Rights System

It may be possible to find settings where all aspects of water allocation, including (formal or informal)[5] rights of use, are defined at the local level among basin stakeholders, but it is more likely that at least some aspects of water allocation are defined within a context of national or provincial rules establishing rights of use, affecting the success with which water users agree upon, maintain, and enforce agreements that regulate use or require contributions to collective action.

Quantification of water rights is such a characteristic. Quantified water rights systems assign rights to a particular amount of water use; nonquantified water rights systems may simply define allowed uses or establish ordinal priorities among users. Advantages of quantified rights systems include relative clarity and certainty among users about who may use what, and in the assignment of tariffs or other fees.[6]

---

[5] The expression "formal or informal" is used here to distinguish between rights that are institutionalized to the point of legal cognition and enforceability, and rights that are norms of entitlement or obligation. Use of the word "rights" is not meant to specify the former or exclude the latter. For the purposes of this study, what is important is not the form in which the right is expressed, but its recognition as binding among stakeholders.

[6] Another characteristic related to ease of reaching resource management agreements is whether the water rights system allows carry-over of unused water rights from one time period to another. "Use it or lose it" systems, in which water not used in the current period cannot be claimed or used in a later period, create unhelpful incentives for water users – for example, trying to enlarge their rights (even resorting to cheating or waste) to guard against uncertainty, overinvesting in conveyance capacity and underinvesting in storage capacity, and so on. Systems that assure water users that forgoing some use today will not diminish their future right, or even allow them to store water in one period for use in the next, lower the costs of reaching and sustaining allocation agreements among users.

Adequate Time for Implementation and Adaptation

Longevity of water resource management arrangements may reflect their success, but their success may also depend on their longevity. Time is needed to develop basin-scale institutional arrangements; to experiment with alternatives and engage in trial-and-error learning; to build trust, so water users begin to accept new arrangements and gradually commit to sustaining them; and for resource management actions to be translated into observable and sustained effects on resource conditions. Several successful examples of basin-level institutions, including some in this study, took decades to design and implement. The relationship between time and success is, however, complex and difficult to define, and depends on a delicate balance between adaptability to changing conditions and patience with existing institutions.

1.4.4
Basin-Level Institutional Arrangements

Successful implementation of decentralized water resource management will also depend on features of the basin-level arrangements created by stakeholders and central government officials, such as:

- Presence of basin-level governance institutions
- Clarity of institutional boundaries and match to basin boundaries
- Recognition of subbasin communities of interest
- Availability of forums for information sharing and communication and for conflict resolution.

Presence of Basin-Level Governance Institutions

A prerequisite of successful resource management is governance – arrangements by which stakeholders articulate interests, share information, communicate and bargain, and take collective decisions. Basin-level governance is essential to the ability of water users to operate at multiple levels of action, which is a key to sustained successful resource preservation and efficient use (Ostrom 1990).

Clarity of Institutional Boundaries and Match to Basin Boundaries

This variable refers to whether and to what extent the institutional arrangements for water resource management have boundaries that are clearly defined and reasonably well matched to the basin (Ostrom 1990). Ill-defined or poorly fitted boundaries may impair collective decisionmaking by including individuals or communities who are not actually in the basin, or excluding others who are. All other things being equal, successful implementation of decentralized water resource management is expected to be found in cases where basin-level institutions have clearly defined boundaries that are reasonably well matched to basin boundaries, and where responsibilities of organizations within the basin are clear to the stakeholders.

Recognition of Subbasin Communities of Interest

Although water resources within a basin are interrelated, it is vital to the success of basin-level governance that the interests of subbasin communities of interest are recognized. Perspectives differ, for example, between downstream and upstream users; between those overlying groundwater resources and those who do not; and between municipal, industrial, and agricultural users as to the reliability of water supply. Recognition can involve mere representation (that is, a guaranteed voice in basin decisions), or it can be extended to construction of basin-level decisionmaking arrangements in such a way that communities of interest must reach agreement on resource management decisions. Of course, transaction costs increase as such assurances are issued to each subgroup within a collectivity, so they need to stop short of a counterproductive threshold.

Availability of Forums for Information Sharing and Communication

Especially in water resource management, where there can be many indicators of water resource conditions and management performance, and many different views on the implications of these indicators, forums for information sharing and communication between stakeholders are vital to reducing information asymmetries and promoting cooperation. All other things being equal, successful examples of decentralized water resource management are expected more frequently among cases where there are basin-level forums for information sharing and communication among stakeholders.

Availability of Forums for Conflict Resolution

Disagreements among stakeholders will arise in any natural resource management setting. The success and sustainability of decentralized management efforts therefore depend also on the presence of forums for airing and resolving conflicts.

All of the above factors have guided the analytical process of this study, both with regard to the global survey and the case studies. They do not all affect with equal significance the decentralization of river basin management in each location. The case study approach was pursued in order to examine closely the processes of institutional change as well as the current situation. Institutional analysis in a case study setting consists largely of determining which institutional factors in what combination appear to have been linked to outcomes. Furthermore, many of the variables listed above have subjective components, so it has been essential in these case studies to interview several individuals with a variety of perspectives.

1.5
Plan of the Book

The remainder of this book presents the results of this study, particularly the application of the analytical framework through eight basin cases and the global survey of river basin organizations.

Part I
River Basin Management at the Lowest Appropriate Level:
When and Why Does It Work in Practice?

This section comprises this chapter plus Chaps. 2 and 3. It introduces the study and its key findings and policy recommendations; presents the chapter with the comparative analysis of the case study basins; and provides the results of the global river basin organization survey and the statistical analysis.

Part II
The Case Studies

In this section, comprising Chaps. 4 through 11, each of the eight case studies is presented in detail for the interested reader.

Part III
Conclusion

A concluding chapter points towards four main conclusions, plus key policy implications of this study and directions for further research.

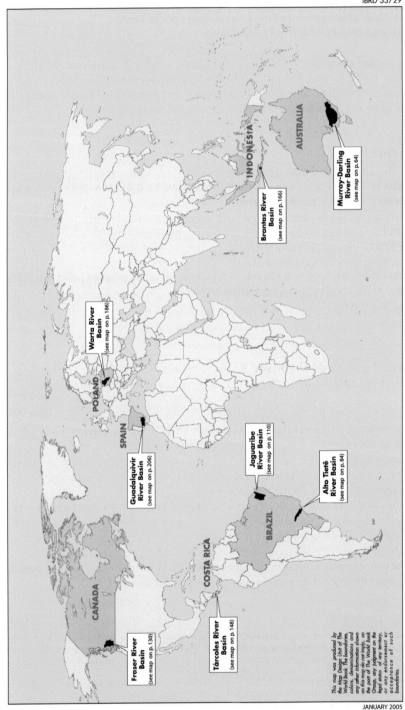

Warta River
Basin
(see map on p.186)

POLAND

SPAIN

Guadalquivir
River Basin
(see map on p.206)

Brantas River
Basin
(see map on p.166)

INDONESIA

AUSTRALIA

Murray-Darling
River Basin
(see map on p.64)

Jaguaribe
River Basin
(see map on p.110)

Alto Tietê
River Basin
(see map on p.84)

BRAZIL

COSTA RICA

Tárcoles River
Basin
(see map on p.148)

CANADA

Fraser River
Basin
(see map on p.130)

This map was produced by
the Map Design Unit of The
World Bank. The boundaries,
colors, denominations and
any other information shown
on this map do not imply, on
the part of The World Bank
Group, any judgment on the
legal status of any territory,
or any endorsement or
acceptance of such
boundaries.

Chapter 2

Comparative Analysis of Case Studies

W. Blomquist · A. Dinar · K. E. Kemper

As noted in Chap. 1, this study of decentralized river basin management posed several questions. What factors might affect the likelihood of stakeholder involvement really contributing to effective basin-level resource management? Are efforts to conduct integrated water resource management at the basin level able to succeed? How can stakeholder involvement and effective resource management at the basin level be sustained over time and changing conditions? What factors might account for the longevity of decentralized arrangements in some cases and their demise in others?

The eight river basin case studies were undertaken to pursue answers to questions such as these. This chapter discusses the eight cases with respect to those questions, while highlighting other factors that emerged as important during the course of the study.

2.1
Characteristics of the River Basin Organizations, Decentralization Reforms, and Approaches to Stakeholder Involvement

The cases studied for this project varied in a number of ways, beginning with the water resource problems that water users and policymakers were trying to address. Table 2.1 presents the three or four most important water management challenges in the basins, listed in order of decreasing priority or severity for each case. In some basins, water supply problems received greatest emphasis in the design and operation of management efforts; in others, water quality.

The cases also differed as regards a variable viewed as potentially significant in the analytical framework presented in Chap. 1: whether the development of institutional arrangements for basin-level water management was initiated through central government reform, or through the efforts of stakeholders. Table 2.2 distinguishes between the origins of the management reforms in the eight cases (though admittedly, sorting them into only two categories compresses subtle differences among the cases).[1]

---

[1] In the Brazilian cases (Jaguaribe and Alto Tietê), because of the federal nature of Brazil's governmental structure, action initiated by state governments would correspond with a top-down initiative. The state-initiated actions in the Brazil cases involved a state government directing the creation of a basin management organization within its borders in much the same fashion as a national government might create one in a unitary system. By contrast, the origin of institutions in the Murray-Darling case started with the states *in their role as basin stakeholders* contesting over supply allocation from an interstate river. Thus for the purposes of this book state-initiated action in that case is classified as stakeholder-initiated action.

Table 2.1. Principal water management problems faced in each basin

Basin	Principal water management problems
Alto Tietê (Brazil)	Pollution, urban development in headwaters, scarcity, flooding
Brantas (Indonesia)	Pollution, flooding, seasonal water scarcity
Fraser (Canada)	Flooding, pollution, intersectoral conflict
Guadalquivir (Spain)	Scarcity and water allocation, drought exposure, flooding, pollution
Jaguaribe (Brazil)	Scarcity and water allocation, water storage, drought exposure, intersectoral conflict
Murray-Darling (Australia)	Scarcity and water allocation, intersectoral conflict, river ecology
Tárcoles (Costa Rica)	Pollution, flooding, erosion
Warta (Poland)	Pollution, flooding, seasonal water scarcity, drought exposure

Table 2.2. Top-down and bottom-up creation of basin organization

Central-government initiated	Stakeholder initiated
Brantas	Fraser
Guadalquivir	Murray-Darling
Warta	Tárcoles
Jaguaribe (national to state, then state to local)	
Alto Tietê (state initiated, with parallel strong stakeholder support)	

A central government might initiate the development of basin management arrangements as a singular act applying to one basin, or as part of a broader water policy reform – for example, creating basin organizations throughout the country. These distinctions are presented in Table 2.3. Not surprisingly, the three stakeholder-initiated cases from Table 2.2 appear in Table 2.3 as cases where creation of basin management arrangements was specific to that basin.

During the study it became apparent that in some cases supranational organizations such as the World Bank and the European Union had been influential in the development or modification of basin management programs or institutions (Table 2.4). European Union policies have influenced water resource and river basin management in Spain and Poland. World Bank promotion of integrated water resource management influenced the development of water management organizations in Ceará state in Brazil (the Jaguaribe case) and the continuity of the basin management corporation approach in Indonesia (the Brantas case). A World Bank-financed project in the state of São Paulo supported the development of legislation that would have influence on the instruments for river basin management in the Alto Tietê basin and on its institutions, such as the

Table 2.3. Basin management reform part of broader decentralization reform?

Yes	No
Brantas[a]	Fraser
Guadalquivir[b]	Murray-Darling
Warta	Tárcoles
Jaguaribe	
Alto Tietê	

[a] The decentralization reform in Indonesia started only after the process started in the Brantas basin. Nevertheless, the broader process has had a major influence in transforming the deconcentration process to a real decentralization process in the Brantas basin.

[b] Initial creation of the basin management agency in the Guadalquivir case occurred in concert with the central government's creation of a number of such agencies in 1926–1927. Significant reorganization of the basin agency and changes in its responsibilities took place in 1985 and 1987 as part of a national water policy reform.

Table 2.4. Presence of supranational influence on creation of basin institutions

Supranational influence present	Supranational influence not present
Brantas (World Bank)	Murray-Darling
Guadalquivir (European Union)	Fraser
Warta (European Union)	
Jaguaribe (World Bank)	
Tárcoles (Inter-American Development Bank)	
Alto Tietê (World Bank)	

Headwaters Protection Law and the Water Pricing Law. The Inter-American Development Bank supported the Tárcoles Commission in Costa Rica.

Basin-scale organizations have been created in each of the eight cases, but they differ in structure and type, as summarized in Table 2.5. Two of the cases featured state companies, two involved central government agencies operating within nationally defined basin boundaries, and the other four were unique variations.

Because integrated water resource management at the river basin level could involve a range of responsibilities and activities, it is not surprising that the cases studied differed in the functions they perform, including authority to allocate water to users; responsibility for water quality; setting and collecting water tariffs; and operation of physical infrastructure. Table 2.6 indicates which among six types of

Table 2.5. Nature of basin organization

Basin	Nature of organization
Alto Tietê	River basin committee supported by a river basin agency
Brantas	State company, under supervision of national water resource agency
Fraser	Nongovernmental organization
Guadalquivir	Central government agency at basin scale
Jaguaribe	River basin committees and commissions supported by the state Water Resource Management Company (COGERH)
Murray-Darling	Intergovernmental basin commission, with a self-financed division for operating infrastructure on Murray River stem
Tárcoles	Quasi-governmental river basin commission recognized by central government
Warta	Central government agency at basin scale

Table 2.6. Basin organization responsibilities

Basin	Planning and/or coordination	Infrastructure operation and maintenance	Licensing water uses/ allocating water supply	Water quality monitoring	Consultation on land use or new water use/discharge permits	Setting/ collecting water charges
Alto Tietê	✓				✓	
Brantas	✓	✓	✓			✓
Fraser	✓					
Guadalquivir	✓	✓	✓	✓		✓
Jaguaribe	✓	✓	✓	✓		✓
Murray-Darling	✓	✓		✓		
Tárcoles	✓					
Warta	✓	✓			✓	

responsibilities were performed by the basin-scale organizations studied in the eight cases. The only function performed by all was planning and coordination – all developed basin management plans or coordinated activities among multiple governmental and nongovernmental entities present within the basins.

As expected, the eight cases also used different means of organizing stakeholder participation in basin management decisions, summarized in Table 2.7. Two cases, Brantas and Warta, had no established stakeholder organization during the time of the study, although one (Brantas) had a program of outreach and communication between basin agency personnel and various individual stakeholders and the other

Table 2.7. Organizational structure of stakeholder involvement

Basin	Stakeholder involvement
Alto Tietê	Basin committee and five regional subcommittees, with representation from civil society, state and local government
Brantas	No stakeholder advisory group or organization (but different government entities representing different user sectors); basin organization has stakeholder outreach program for communication; new law provides for establishment of an advisory council
Fraser	Multisector council with representation from basin subregions, water use sectors, and all levels of government; regional councils
Guadalquivir	Water Users Assembly, and several basin commissions and councils with seats for water user groups, civil society, and central government
Jaguaribe	Numerous user commissions at reservoir and valley scales, plus subbasin committees
Murray-Darling	Community Advisory Committee with representation from basin subregions, water use sectors, and state and local governments
Tárcoles	Multisector commission with representation from civil society, national and local government
Warta	No stakeholder advisory group or organization at time of study; one in process of being established

(Warta) is now developing a regional water management council under the direction of a national law. Other cases, such as the Alto Tietê and Fraser basins, had elaborate and multiscale structures. Jaguaribe had numerous subbasin user committees and commissions, but only COGERH, which provides the technical support to these, operates at the basin scale, and Murray-Darling has a basinwide Community Advisory Committee but not subbasin ones. The Guadalquivir and Tárcoles cases have representative structures incorporating a variety of stakeholders.

The financing of the basin organizations is similarly varied. As shown in Table 2.8, three (Alto Tietê, Tárcoles, and Warta) rely solely on central government budget allocations at the time of writing, though the basin agency and committee in Alto Tietê are supposed to have revenue from water charges in the future. Three others (Brantas, Guadalquivir, and Murray-Darling) enjoy a combination of central government support and water user charges. One (Jaguaribe) is funded entirely by water user charges, although those are collected by the state water management agency from users outside as well as within the basin and then reallocated to the basin. The nongovernmental Fraser Basin Council lacks the authority to levy taxes or charges on water use, and instead receives annual financial support from governments and project funds from a variety of sources.

Undoubtedly more factors could be used to characterize and contrast the eight cases, but Tables 2.1–2.8 provide an overview of similarities and differences. In a comparative case study it is not necessary to have a representative sample, but it is important to have substantial differences among cases on variables of potential interest. These eight cases plainly do meet that requirement.

Table 2.8. Financial arrangements associated with basin organization

Basin	Financial arrangements
Alto Tietê	Currently, funding received from earmarked State Water Resources Fund. In the future, water user charges will cofund basin organization activities
Brantas	Fees from some users (industry, hydropower, and municipalities) cover most operation and maintenance activities. Central government funds staff salaries, flood control and major infrastructure projects
Fraser	Federal, provincial, and local government annual contributions, plus funds collected from sponsors of Fraser Basin Council projects
Guadalquivir	Funding from central government plus fees collected from water users and water suppliers in the basin
Jaguaribe	Water charges collected by COGERH, but cross-subsidization from other basins, especially the urbanized basin of the Greater Fortaleza region
Murray-Darling	Basin commission funded by national government and states; management of Murray River infrastructure by fees on bulk water users
Tárcoles	Funding currently coming from central government; past support provided by foundations and Inter-American Development Bank
Warta	Budget allocation from central government

2.2
The Basin Cases and the Analytical Framework

Tables 2.9–2.12 condense findings about the eight basins with respect to the four categories of variables in the analytical framework – contextual factors and initial conditions, characteristics of the decentralization process, central-local relationships and capacities, and basin-level institutional arrangements.

2.3
Questions, Findings, and Conclusions

In this section, relevant questions are posed in order to draw some conclusions across the cases about important factors that were related to the degree of success in gaining stakeholder involvement, developing institutions at the river basin level for integrated water resource management, and addressing water resource problems. Table 2.13 provides a condensed statement of outcomes of decentralization and water management reforms for each case.

Was the Active Involvement of Stakeholders Secured?

In most cases the answer is yes, although there are exceptions and variations. Representation of diverse groups of stakeholders, regular and sustained opportunities for interaction, an ambitious agenda of basin management issues, and direct connection of basin management activities with matters relevant to people's livelihoods and local communities contributed to active involvement in several of the cases observed.

Table 2.7 illustrates the range of stakeholder involvement. The river basin organizations of the Warta (Poland) and Brantas (Indonesia) basins lacked organized structures for stakeholder involvement. In the Guadalquivir basin (Spain) the number of stakeholder organizations has expanded, though some operated more efficiently than others. In the Tárcoles basin (Costa Rica), stakeholder involvement was quite active in the 1990s as the basin commission was getting started. Local stakeholders participated actively with state government personnel in establishing basin and subbasin management organizations in the Alto Tietê basin in Brazil. In the Jaguaribe basin in Brazil, participation takes place along the river valley where the key infrastructure is located and around strategic multiyear reservoirs. A high level of stakeholder engagement has been achieved in the Fraser and Murray-Darling cases.

What Factors Appear to Be Related to Successful Start-Up of River Basin Organizations?

The commitment of governmental support to the creation of stakeholder organizations was a positive factor at the outset of the Tárcoles, Fraser, Jaguaribe, and Alto Tietê experiences. The presence or prospect of valuable infrastructure investments became a point of stakeholder interest from the beginning of the Guadalquivir and Murray-Darling cases. The absence of significant cultural conflicts among basin stakeholders in most of the cases helped too.

Beyond the analytical framework, there were other factors that proved to be important. One was the presence of a champion, an influential individual who drew attention to basin problems and conditions and gave impetus to the reform process, as in the Brantas, Jaguaribe, Fraser, and Tárcoles cases. Another was the stimulus provided by the presence of severe water resource problems – in short, in several of the cases water resource conditions deteriorated to a point where stakeholders found some form of engagement nearly unavoidable. A third factor was the influence of supranational entities such as the World Bank or the European Union and their support for the organization of water resource management at the river basin scale.

Has the Active Involvement of Stakeholders Been Sustained over Time?

The answer to this is naturally complicated by variations in stakeholder involvement[2] and lengths of time basin management has been undertaken. Where sustained active involvement occurred, it appeared to be connected with stakeholder perceptions that the basin management organizations were engaged in important issues, were making (or had made) a positive difference in basin conditions, had

---

[2] Cases such as the Warta and Brantas basins, for instance, which had not yet institutionalized particular mechanisms for stakeholder involvement by the time of the study, might be categorized as "not applicable" with respect to this question.

Table 2.9. Summary of contextual factors and initial conditions

Basin	Contextual factors and initial conditions
Alto Tietê	Despite a favorable socioeconomic context and the initial distribution of resources among basin stakeholders seeming to favor reform, the political will to advance the changes has proved insufficient. The political and environmental complexities of the Alto Tietê basin seem to make it particularly difficult to implement practices involving integration and participation in decisionmaking.
Brantas	Weak institutions, a poor legal and regulatory framework, ineffective bureaucracy, and endemic corruption led, in the aftermath of the 1997 Asian financial crisis, to policy and institutional reform assisted by external donors and based upon macroeconomic management and stabilization as well as governmental reform. Since 1999, the country has gradually recovered macroeconomic and political stability.
Fraser	Economic and social conditions are conducive to successful river basin management. Most land and water resources in the Fraser basin are held by the province of British Columbia, and used by private individuals under lease arrangements with the government. Therefore, no single interest or sector of basin users had legal immunity for their claims or titles to resource use.
Guadalquivir	While ethnic, religious, or class divisions in Andalusian society do not seem to have driven water management issues, economic development seems to have had a notable effect. The Guadalquivir basin was poorer and more rural than most of the rest of the country, and these conditions contributed to an emphasis on the expansion and protection of irrigated agriculture.
Jaguaribe	Local conditions in the Jaguaribe basin appeared to be unfavorable for decentralized water resource management. The basin is relatively poor; participatory water management ran contrary to the prevailing political culture; and Ceará state had one of the most entrenched oligarchies in the Northeast. Factors favoring reform included a national transition towards democracy, and increased promotion of integrated water resource management by the technical water resource community.
Murray-Darling	Factors favoring the development of basin management institutions included the semiarid climate, which makes water issues significant enough to stimulate action, and the relative wealth and homogeneity of its population, presenting few barriers to such action. The initial distribution of resources among basin stakeholders has clearly favored irrigators in the basin, who account for more than 90 percent of water diversions.
Tárcoles	The Tárcoles basin is by far the most economically developed in the country and there do not appear to be substantial cultural or religious differences that would inhibit cooperation. Financial resources are, however, a limiting factor, and there seems to be overall political reluctance towards decentralization to lower levels of decisionmaking.
Warta	The Warta basin does not feature significant cultural, religious, ethnic, or other divisions. Poland's economic conditions have led to financial constraints on the government sector, limiting its ability to provide either central funding or revenue autonomy adequate to the tasks of integrated water resource management at the basin scale.

consistent governmental support, and were operating regularly and frequently. These perceptions were in evidence in the Fraser and Murray-Darling cases, at the subbasin levels

Table 2.10. Summary of decentralization process characteristics

Basin	Decentralization process characteristics
Alto Tietê	The decentralization process in the Alto Tietê basin is marked by two distinct processes: decentralization from the state to the basin level, which occurred with the creation of the Alto Tietê Committee in 1994 and, more recently, its water agency; and further decentralization within the basin in 1997–1998, which resulted in five subcommittees at lower territorial levels. Neither state government not local stakeholders question either the need to create complementary deliberative bodies at lower levels or the fact that basin management participants should be allowed to create and modify institutional arrangements according to their needs and circumstances. However, there is no agreement about the extent of this decentralization.
Brantas	The decentralization process has generally been government led and top down. Passage of the new Water Law, in March 2004, signals central government commitment to continued reform of the water resource sector in accordance with the agreed action plan developed under the World Bank-assisted Water Sector Adjustment Loan.
Fraser	Development of the Fraser Basin Council grew out of previous but less comprehensive intergovernmental projects. The construction of basin-scale institutional arrangements in the Fraser basin has been a matter of integrating already decentralized organizations and jurisdictions. The extent of central government recognition of local-level basin governance has been positive.
Guadalquivir	The confederaciones hidrografica (CHs) are central government agencies with representative components. While basin-scale institutions enjoy the recognition of central government officials as legitimate water resource management entities, such recognition has not been accompanied by an extensive devolution of authority to basin-scale institutions.
Jaguaribe	Decentralization in the Jaguaribe basin was marked by two distinct stages: decentralization from federal to state level, including the creation of a state water resource agency (COGERH); and decentralization from state to local level, which occurred through the creation of deliberative bodies at the river basin and lower territorial levels. Some state-level authorities appear distrustful of the participatory decisionmaking bodies created, despite the advances made in the last decade.
Murray-Darling	Primary decisionmaking authority predominantly rests at the subbasin level with the state governments. Over time, and with the cooperation and consent of the national government, the states have constructed intergovernmental arrangements to control and operate Murray River flows and then to address other issues. The process has been as much a matter of integration as decentralization. Central-level recognition of basin governance and management has been complete and consistent.
Tárcoles	The Commission for the Río Grande de Tárcoles Basin (CRGT) was essentially a municipal initiative and took an active leadership role. The central government partially devolved authority to the CRGT and was supportive of its efforts, but there was never full recognition of the CRGT's authority to manage the basin. Since 1998 the central government has neither pushed the devolution forward nor terminated the commission.
Warta	Creation of river basin agencies and reform of water policy were attempted in the same decade (1990s) as the overall democratization of Polish government, resulting in some uneven progress. Significant responsibilities for water resource planning and management have been spread across basin and subbasin entities, and water law reform took several years longer than originally envisioned. The central government has maintained commitment to decentralization, but financial resources have been a problem.

in the Alto Tietê and Jaguaribe cases, and with respect to irrigation communities in the Guadalquivir basin.

Stakeholder involvement has waned during the past five years in the Tárcoles case, as well as in the Jaguaribe and Alto Tietê cases, and is lower among

Table 2.11. Summary of central-local relationships and capacities

Basin	Central-local relationships and capacities
Alto Tietê	The advances in state water management capacity have been considerable and in some cases crucial for the survival of the basin committees in this transitory phase. However, tensions and problems exist between the central authorities and the local bodies and basin committees are not always effective. Indeed, the São Paulo water resource management system as a whole is beginning to show signs of breakdown in the face of the state government's incapacity to make it fully operational, especially by implementing bulk water charges. Water reform in São Paulo seems to need much more time. Considering that the reform process is almost 15 years old, it is becoming clear that transaction costs are very high in terms of time and money – so much so that the pioneer state in water reform has begun to lag behind others.
Brantas	Much power still resides with central government ministries for planning and policymaking. The authority to oversee the management and functioning of the Brantas River Basin Management Corporation (PJT I) lies with the center through the Ministry of Settlement and Regional Infrastructure (Kimpraswil), with the Ministry of Finance exercising a fiscal oversight role. The supervisory board of PJT I does not have a basin-level stakeholder institution to work with.
Fraser	Financial resources and autonomy of the council are relatively strong. The council members have, through the Fraser Basin Society and the council's own bylaws, the demonstrated ability to create and modify the institutional arrangements with which they work. The arrangements governing rights to water and land use allow for considerable management flexibility, though the control of groundwater resources is weak and represents a vulnerability in terms of the overall basin sustainability effort.
Guadalquivir	Entities such as CH Guadalquivir construct basin-level plans, but these plans must be submitted for national approval and be consistent with the national water plan. CHs collect and maintain revenue of their own for some of the services they provide but also rely on central government funding for functions determined by central government officials. CHs have several advisory bodies composed of stakeholder representatives, but several of those councils also have central government representatives and the CH president is still a central government designee. The central government remains free to alter CH governance structure processes with as much or as little stakeholder consultation as it chooses.
Jaguaribe	The devolution of some federal authority over the management and control of reservoirs to Ceará state has been effective, since COGERH has developed substantial technical, administrative, and financial management capacities. But other aspects of water management decentralization remain underdeveloped, for example groundwater management, and the basin committees lack effective technical, administrative, and financial support. Ceará has centralized water charging at the state level, and the committees have been limited to information dissemination, capacity building among local actors, and the resolution of water use conflicts. There has been resistance within COGERH to giving decentralized bodies greater power over water management, and only the subbasin committees have been legally created.
Murray-Darling	Generally, central-local relationships and capacities are favorable to integrated water resource management. The basin management participants have the ability to create and modify institutional arrangements. There is considerable experience at the local and state levels with self-governance and service provision. One hindrance to basin-level integrated water management is the system of water rights. A second hindrance is that the organizations in the basin most directly associated with integrated resource management (for example the subbasin catchment management authorities) have virtually no financial resources of their own and are dependent on government funding.

nonirrigation stakeholders in the Guadalquivir case, for reasons converse to those above, particularly perceived inconsistency of government support and lack of encouragement for and communication between certain stakeholders.

Table 2.11. *Continued*

Basin	Central-local relationships and capacities
Tárcoles	Financial resources for the basin management effort have always been limited, and the CRGT has never had its own revenue stream. This has severely limited the commission's ability to evolve into something more than a meeting place. Cantons and municipalities do perform a number of functions, so there appears to be local-level experience with self-governance and service provision, rather than an excessive centralization of public services. The ability of any river basin commission in Costa Rica to develop and implement effective water supply management policies is likely to be hampered also by the weak framework of water rights allocation.
Warta	Overall, the water law changes in 1997 and 2001, and the merger with the centrally appointed district directorates of water management (DDWMs) in 1999, have given the regional water management authorities (RWMAs) more responsibilities but not additional sources of revenue. The water rights system is conducive to integrated water resource management, and a well-developed structure is in place. Permits for water use and water discharge are limited in time and quantity, and fees associated with nonpermitted actions or with permit violations provide incentives to users and also a revenue source for environmental improvement projects.

Was Stakeholder Involvement Linked in a Substantive Way to Resource Management Decisionmaking (As Distinct from Mere Stakeholder Consultation)?

In most cases, the answer to this question was yes. Relatively new as well as long-lived river basin organizations had engaged stakeholders in substantive basin management decisions. Stakeholder involvement was more common with respect to basin planning, water supply allocation, and infrastructure operation, and less common with respect to setting water charges, collecting fees, flood control, monitoring basin conditions, altering land uses, or infrastructure construction decisions.

In the Guadalquivir basin, operations boards that include stakeholders and agency staff make reservoir management decisions, for example water allocation for irrigation. In the Jaguaribe basin, user committees at the local reservoir scale have a strong say in annual water storage and releases. In the Alto Tietê basin, the river basin committee and subbasin committees are involved in designating headwater protection areas. Catchment management authorities in the Murray-Darling basin stakeholder organizations are involved in land use change monitoring, allocation of irrigation water, and management of dams. Stakeholders in the Tárcoles basin were actively involved during the 1990s in programs of reforestation and reduction in contaminant discharges. The nongovernmental Fraser Basin Council forms partnerships with other governmental and nongovernmental entities in the basin for specific projects. Stakeholder involvement in basin management in the Brantas and Warta basin cases has been more extensive with respect to the review of basin plans prepared by agency staff than in the making of basin management decisions or the operation of projects.

Did Stakeholder Involvement Translate into More Effective Resource Management?

Stakeholder involvement and performance improvements have gone hand in hand in reducing exposure to flooding and better management of releases from water storage

Table 2.12. Summary of basin-level institutional arrangements

Basin	Basin-level institutional arrangements
Alto Tietê	All of the new institutions defined in the Water Law have been formally implemented; however, their operation is still marked by imprecision and institutional overlap, largely as a result of the varying performance of the public representatives. Gaining influence over state programs is the main challenge for all basin committees in Brazil, especially for those with little or no capacity for implementing a water pricing system. The subcommittees are generally considered more dynamic, more effective, and more important than the main committee. The most important role of the subcommittees is to deal with making water resource protection and urban expansion compatible through the implementation of the state Headwaters Protection Law of 1997. The Headwaters Protection Law recognizes that simple prohibition and policing measures for protecting strategic water supply sources have had perverse effects. The new approach is a dramatic departure from São Paulo's traditional sectoral approach to water quantity and quality, which separated the management of water from its environmental aspects, especially water pollution and land use.
Brantas	PJT I is a state-owned company with clearly delineated management responsibilities and a profit motive. This arrangement has permitted the company to focus on the river basin as the management unit and on management rather than development and construction, and has also endowed the company with credibility that the funds it receives from water users will be reapplied in the basin. PJT I shares responsibility with and is ultimately subservient to other political entities at provincial and national levels with respect to the development of water resource management policy.
Fraser	The Fraser Basin Council was designed for stakeholder information sharing and communication and has emerged as the paramount deliberative body in the basin. In its capacity as a nongovernmental organization funded through a nonprofit society, however, the council must often hand off projects to other entities for implementation. At times even the council members are not entirely clear what actions are within the council's scope.
Guadalquivir	While geographic boundaries of the river basin and CH Guadalquivir fit well, institutional boundaries are less clear. The basin-level institutional arrangements do recognize subwatershed communities of interest within the basin. However, only irrigation user communities have formal recognition in both national law and the CH organizational structure. Basin-level institutional arrangements are structured to provide forums for information sharing and communication among basin stakeholders and between stakeholders and CH staff. The effectiveness of these structures varies.
Jaguaribe	The prestige attached to water allocation in the basin has led some representatives of subbasin committees to argue that the Jaguaribe-Banabuiú Valleys Commission should be dismantled, with the transfer of its responsibilities to the committees. All actors in the basin have, however, shown support for the 36 reservoir user commissions, whose decisions have only localized impact involving few transaction costs. Despite the current uncertainties concerning institutional boundaries, both user commissions and basin committees have been promoting the resolution of water use conflicts, with the support of COGERH.
Murray-Darling	There are basin-level governance organizations and subbasin organizations, each with firm recognition and considerable support from the state and Commonwealth governments. The states themselves are recognized as communities of interest within the river basin, as are a number of stakeholder communities represented on the Community Advisory Committee. Basin users and policymakers have an array of means by which to negotiate and enter into agreements for committing and combining resources for projects and programs to improve basin conditions, which are regularly monitored. Clarity of institutional boundaries has been somewhat reduced by the introduction of the relatively new catchment management bodies.

reservoirs in the Brantas, Guadalquivir, Jaguaribe, and Murray-Darling cases; in the reduction in the rate of deforestation in the Tárcoles basin; in improved treatment of

Table 2.12. *Continued*

Basin	Basin-level institutional arrangements
Tárcoles	Once it lost its central government support and its dynamic initial leadership, the CRGT's status and composition left it vulnerable to becoming more of a discussion forum than a governing body. The prevailing view that water has to be managed by its uses (for example drinking, irrigation, or hydropower) rather than in an integrated fashion has been reflected and reinforced by Costa Rican laws. There is considerable fragmentation and territorialism among agencies and institutes at the central government and municipality levels.
Warta	Management of the Warta basin is substantially dispersed and polycentric. This federal approach, with responsibilities spread across levels and units of government, allows for the recognition of subbasin communities of interest, and provides overlapping layers of monitoring and enforcement of water management regulations. The federated structure does not, however, lend itself to clarity of institutional boundaries or a close matching of jurisdictional boundaries to basin boundaries.

industrial wastewater and reduced use of the river for waste discharge in the Fraser basin; and in headwater area protection in the Alto Tietê basin. Measurable improvements to wastewater treatment in the Warta basin have resulted from the financial investments of the provincial funds for environmental protection and water management.

On the other hand, stakeholder involvement can perpetuate impediments to improved water resource management. For example, agricultural users in the Brantas basin and public hydropower producers in the Tárcoles basin remain exempt from tariff requirements that apply to other users. In these instances asymmetry in the political influence of certain water use sectors threatens the improvements supposed to arise from integrated water resource management and stakeholder involvement.

Although the degree of success and longevity of basin management efforts may vary across cases, it can be concluded that the social and institutional capital in all eight cases is richer due to the actions that have been taken, and they are in a better position to meet future water management challenges. None of the cases appears to be in worsened condition as a result of stakeholder involvement and integrated water resource management initiatives or the effort to develop basin-scale institutional arrangements.

Were Improvements in Resource Management Sustained over Time and As Conditions Changed?

Despite improvements, significant water resource management problems remain in all of the cases studied. In several cases, this was due to changing conditions to which management entities have not adapted fully.

The long-lived management regime of the Murray-Darling basin is wrestling with more recently recognized problems of dryland salinity and deteriorated river ecology. The equally long-lived Guadalquivir basin agency has yet to fully cope with its enlarged responsibilities for water licensing and demand management. In the Warta case several factors, including jurisdictional gaps between the issuance of permits and the enforcement of compliance, have contributed to the rise of seasonal water scarcity in

Table 2.13. Summary of reform outcomes and lessons

Basin	Reform outcomes and lessons
Alto Tietê	Several factors have hindered the Alto Tietê Committee taking the lead in coordinating water management, including lack of political will and the difficulty for stakeholders to identify common interests and come to grips with the technical and financial challenges of such a densely populated and industrialized region. The use of permits and charges for water use has been authorized but implementation has stalled, and therefore the river basin committee and agency lack a reliable and autonomous revenue source; budget support from the state is adequate only to fund a few staff and cannot support any projects or programs at the basin level. While the main committee still tries to define its roles the subcommittees have lacked leadership and not functioned effectively. By default, river basin management has devolved to the level of individual agencies, and decisionmaking remains fragmented. There has been significant mobilization around water issues, but the process has yet to reveal measurable results in improving water quality or rationalizing water use.
Brantas	The state company, PJT I, focuses on being a reliable provider for the tasks it has authority over: water allocation and supply, and flood control. It has implemented an acceptable system of water allocation and management and a reliable flood forecasting system, and appears to maintain major infrastructure in satisfactory condition. It has a neutral and competent image, aided by its focus on a few major activities. Managing water quality and river ecology remain the responsibility of many entities, however, and given their growing importance the authority and coordination to address them need to be improved. Despite the legal status of PJT I as a state corporation its autonomy remains limited, as its supervisory board is dominated by central government representatives and its budget is still heavily subsidized. PJT I lacks authority to set tariffs, cannot collect user fees from irrigators at all, and does not issue water use permits.
Fraser	The nongovernmental organization approach has provided a means of crossing jurisdictional boundaries within the federal system. It has succeeded in allowing the integration of indigenous communities and private stakeholders; it has provided a good forum for information generation and sharing; and it has succeeded in preserving a reputation for objectivity and in building a more diverse financial base. One key to this success has been the ability of the Fraser Basin Council, as a nongovernmental organization, to take on an extremely broad range of issues, though it must rely on other agencies for the implementation of its plans and programs, and its reliance upon consensus occasionally slows decisionmaking.
Guadalquivir	Considerable institutional change has occurred during the long history of the basin agency CH Guadalquivir. The movement of Spanish water policy away from an emphasis on supply augmentation and towards integrated water resource management has added new responsibilities to basin agencies, and has resulted in representation of a broader range of stakeholders on agency boards and commissions. Nevertheless, the basin agency is still perceived by some as focusing mainly on supply augmentation while newer functions – water licensing, demand management – have been performed with less positive results. Despite substantial investments in supply and storage facilities Guadalquivir's water deficit has not been erased. Systematic efforts to reduce nonpoint pollution are still lacking, and a considerable amount of agricultural water use remains comparatively inefficient.
Jaguaribe	The devolution of management of federal reservoirs to Ceará state has been effective; devolution from state to local level has been more partial. Long effort to solve scarcity problems by building reservoirs was only partially successful, which has contributed to the recent emphasis on improving allocation of existing supplies. The creation of subbasin committees and user commissions has increased participation, but stakeholder involvement has been limited largely to negotiating water allocation and resolving conflict. Basin committees in Ceará do not have their own executive structures (for example basin agencies) and have fewer powers than the state over issues such as bulk water pricing. The financial resources of the committees and user commissions are dependent on contributions from state government and from their members, and thus remain insecure. At the state level, though, bulk water pricing has allowed the state company to achieve financial stability for its infrastructure operations. While those state company operations bring funding into the basin, the state company also exports water from the basin for use in the Fortaleza region.

Table 2.13. *Continued*

Basin	Reform outcomes and lessons
Murray-Darling	Murray-Darling basin water resource management's successes in gaining intergovernmental cooperation and instituting mechanisms for stakeholder participation are considerable, and arrangements for devolution of authority, stakeholder participation, and financial self-sufficiency have been generally successful. Much activity is carried out at regional level in local offices with almost complete authority for policy implementation. Management and operation of dams and irrigation schemes has been transferred to entities designed for localized day-to-day management. Both urban and rural water supply infrastructure schemes are self-financed for operation and maintenance. All levels of water management feature stakeholder advisory groups.
Tárcoles	The Tárcoles Commission was for a period in the 1990s able to initiate important basin improvement activities, including a reduction in agribusiness contamination of water and slowing of land degradation through reforestation. However, changes of leadership and increased central government control have been associated with a decline of the commission's activity and stakeholder participation. The Tárcoles basin still lacks sewage treatment, and river water quality conditions continue to worsen. As regards water rights, the current concession system does not cover groundwater use, and the tariff system for agricultural water continues to base fees on cultivated area rather than water use, providing little incentive for efficiency. Groundwater use appears to lack control.
Warta	The Warta basin illustrates how much institutional creation and policy reform can be accomplished in a relatively short period when central government sustains a commitment to decentralization and integrated water resource management. Fifteen years ago Poland lacked a rational system of water tariffs, wastewater discharge controls, water resource planning, or river basin-scale organizations. These are in place now, along with bodies at the national, provincial, and local levels for funding water quality improvements and other environmental protection projects. However, institutional boundaries have not always been clear, and in some respects the establishment of the river basin agencies was premature. Water quality remains a great challenge; funding for water storage facilities, such as small reservoirs, has been inadequate; and the risk of flooding has not been eliminated.

portions of the basin. In the Alto Tietê case, the main problems of São Paulo's water supply have not been sufficiently tackled and keep growing.

In most of the cases, the principal water resource problems that gave rise to the establishment of basin and subbasin organizations are being addressed, and improvements have occurred. It is too early to tell, however, whether the arrangements developed for other problems and functions will cope as newer problems come to the fore.

What Factors Appear to Be Related to the Longevity of Decentralized Arrangements in Some Cases and Their Demise in Others?

The consistency of central government support for basin management, stakeholder involvement, and water policy reform has emerged as one of the most important factors common to cases with greater levels of success and stakeholder participation, and may be as important as magnitude of support over the long run. Magnitude of support matters, too, as can be seen in the negative effects of insufficient central government funding on basin management in the Alto Tietê, Warta, and Tárcoles (post-1998) cases. By contrast, organizations such as the Murray-Darling Basin Commission, the Fraser Basin Council, and CH Guadalquivir have been able to sustain multiyear basin planning and projects thanks to relatively consistent levels of central or other government support for their work.

Financial autonomy is a related factor: most revenue generated from water users in the Guadalquivir, Murray-Darling, and Brantas basins, for example, remains within or is returned to the basin for improvements and operations there. Basins that are wholly dependent on central government allocations (Warta and Alto Tietê, for example) have had a more difficult time establishing their own priorities and undertaking substantial projects. The Jaguaribe case features revenue finding its way into basin projects at the expense of some loss of control over the basin's water supply, part of which is exported to the nearby Fortaleza area.

Other factors apparently encouraging longevity are bottom-up initiation, particularly in settings where there is experience of local autonomy, as in the Fraser and Murray-Darling cases; relatively low levels of cultural conflict among stakeholders; recognition of subbasin communities of interest; incorporation of existing local institutions; and the importance of champions in getting basin management institutions started.

2.4
Implications for Policy and Research

One important finding from these case studies is that, while the level of economic development of the nation and the basin can make it easier or harder to create and sustain basin-level institutions, improvements in water resource management can be and have been realized in a variety of settings, including poorer countries and basins facing severe water problems. Furthermore, improvements can occur fairly early in the life of basin organizations and stakeholder participation initiatives.

Nor do all decisions and activities that contribute to integrated water resource management have to be organized at the basin scale. As has been seen, the lowest appropriate level for some water resource management functions may be a subbasin unit, a local or regional unit of government, or a hybrid unit sometimes referred to as a "social basin" (for example the basin subcommittees in the Alto Tietê case).

In addition, it is vital to recognize that the establishment of participatory and decisionmaking structures involves shifts of power, which can be a controversial and complicating factor. Efforts by the Andalusian regional government to exercise more leadership over basin management in the Guadalquivir case, and the desire of the state company in Indonesia to also take on the pollution control currently the responsibility of provincial government, are only some examples of how jurisdictional and other power-related considerations are likely to arise. The political economy will always play a role in the water sector and is an important factor to be borne in mind.

Finally, decentralization reforms and the establishment of river basin management with active stakeholder involvement are processes that take time, even decades. This is why consistency of support is so important. So too is the ability to adapt and modify basin management arrangements in response to changed situations. Central governments and external organizations wishing to promote integrated water resource management on a river basin scale should be prepared to sustain their commitment to reform, across changes of administration and through good and bad times.

Chapter 3

Determinants of River Basin Management Decentralization:
Motivation, Process, and Performance

A. Dinar · K. E. Kemper · W. Blomquist

3.1
Introduction

In recent years, increased awareness and concerns among both policymakers and water users regarding the state of water and its provision have motivated implementation of various reforms in the water sector, including at the river basin level. As described in the previous chapters, one of the major components of these reforms has been the decentralization of river basin management to the lowest appropriate level. Such reform implies the involvement of different stakeholders in the basin, including water users, in order to achieve more sustainable management of the basin's water resources.

But increased stakeholder involvement requires responses to several important related questions. If such active involvement of stakeholders is secured, how can it be translated into effective resource management and high performance level? What factors might affect the likelihood of stakeholder involvement turning into effective basin-level resource management? If stakeholder involvement is translated into basin-level management, how can the active involvement and effective resource management be sustained over time and changing conditions? What factors might account for the longevity of decentralized arrangements in some cases and their demise in others?

Answers to these and related questions are still unclear. Due to the relative recentness of the approach, and the fact that evaluation of decentralization and performance presents conceptual and empirical difficulties, few analyses of decentralization of river basin management functions have as yet been undertaken (see Mody 2004 for a review of the literature). Also, while those case studies that have been undertaken shed light on the direction of development in river basin decentralization, their individual focus does not permit the identification of generic forces shaping the process of decentralization or factors accounting for its level of success. An analytical framework, developed by Blomquist et al. (2005) and described in Chap. 1, allows identification of these general relationships and patterns and is applied to the principal case studies considered in this book. The framework incorporates the political, institutional, and economic variables and the paths by which they may influence the decentralization outcome. It takes into account initial and contextual conditions and, in recognition of the fact that the lowest appropriate level for integrated river basin management varies between basins, it includes consideration of hydrological, socioeconomic, cultural, and historical conditions in each basin (see also Saleth and Dinar 2004).

The analytical framework is expanded on in this chapter to explain levels of success of decentralization processes across a large set of river basins. The following sections present the hypotheses that will be tested; the process of data collection and variable formation; the results of the analysis and their interpretation; and conclusions and policy implications.

3.2
Development of Hypotheses

This section focuses on developing the analytical framework as the basis for the development of the hypotheses to be tested in a quantitative analysis that will be described below.

3.2.1
Background

Decentralization of decisionmaking is not an aim in itself: it is recommended because experience over the past decades has shown that when decisionmaking is centralized and local conditions are not taken appropriately into account, then account-ability of decisionmakers is weak, and water resource management is inadequate. On the other hand, a process of decentralization that is appropriate to local circumstances can lead to positive outcomes arising from increases in transparency and from stake-holder involvement in decisionmaking, including measures to accord financial self-sufficiency.

As decentralization is a reform process, other processes leading up to or running parallel with it may affect it. These influences are dynamic and are often subject to change as they interact with the reform process itself; they may be external to or lie within the decentralization process. A number of these influences, which may either encourage or inhibit the reform process, have been identified and analyzed in the literature (Blomquist et al. 2005; Saleth and Dinar 2004; Bromley 1989; Ostrom 1990; Ostrom et al. 1994). They include:

- Tendency towards inertia, perhaps due to the transaction costs involved in change, or resulting from path dependency – the tendency of a community or region to continue along the path it is already following (Saleth and Dinar 2004). On the other hand, the institutional environment may encourage experimentation and innovation.
- Level of influence of various stakeholders, determined, for example, by the asymmetries of power, information, or other resource distribution among them, which will influence their perception of the costs and benefits associated with the reform process.
- Social (or otherwise derived) norms of trust and reciprocity, as expressed in the history of past interactions among individuals and their anticipations concerning future interactions, and which will be influenced by cultural or other differences among the individuals who are attempting to coordinate behavior or whose coopera-tion is needed.

- Presence of incentives for stakeholders to act, for example for the government to decentralize, or for water users and other stakeholders to take on responsibilities.
- Principal-agent relationships, including the transparency and enforcement possibilities in contractual agreements between the stakeholders to carry out certain functions.

For the purposes of this analysis of decentralization of river basin management and its performance, these broad categories can be used to assemble a number of variables that might have some degree of influence on the process of decentralization. Empirical hypotheses have been formulated as to how each variable might contribute to the likelihood of successful or unsuccessful decentralization of river basin management. The variables identified are grouped under four main headings:

1. Contextual factors and initial conditions
2. Characteristics of the decentralization process
3. Central-local relationships and capacities
4. Basin-level institutional arrangements

3.2.2
Hypotheses: Analysis of Variables

Contextual Factors and Initial Conditions

The literature on decentralized water resource management indicates that the outcome of decentralization is partly a function of the initial conditions that prevail at the time a decentralization initiative is attempted (path dependency). These initial conditions are elements of the economic, political, and social context of the decentralization effort. Several variables that could capture such conditions are detailed below.

Level of economic development of the nation measures the ability of the government to financially support the decentralization process. Although a decentralization initiative may be undertaken with expectation to reduce the central government's financial outlays for river basin management, the early stages of decentralization may require some additional outlays in order to make the transition; furthermore, some elements of water resource management have the characteristics of public goods and may continue to be provided by central government.

All other things being equal, decentralization initiatives are expected to be more likely to achieve sustainable success where the economic well-being of the nation allows the central government to bear these initial and ongoing costs.

Level of economic development of the river basin region measures the ability of the basin stakeholders to commit financial and other resources necessary to the decentralization process in addition to support from central government. A degree of financial autonomy, for example through some form of water pricing or tariffs, makes financial resources available at the basin level and assists the development and maintenance of institutional arrangements for basin-level management (Cerniglia 2003; Musgrave 1997).

All other things being equal, basins that have a level of economic development that can sustain those resource commitments are more likely to achieve sustainable success in decentralization.

Initial distribution of resources among basin stakeholders is an important contextual factor that may manifest itself in different ways. On the one hand, extreme disparities in resource endowments may imperil decentralization; privileged stakeholders may not want to participate in a process that makes them worse off, and destitute stakeholders may elect not to participate in a process to which they are unable to contribute any resources of their own. On the other hand, some inequality of initial resource endowments may facilitate action by enabling some stakeholders to bear the costs of taking a leadership role (Blomquist 1988; Ostrom 1990).

The relationship between level of inequality of resource endowments and successful decentralization is hypothesized to be quadratic, with greatest positive impact at a certain level of inequality and lower or negative impacts at both lower and higher levels of inequality of resource endowment distribution.

Characteristics of the Decentralization Process

Certain conditions or characteristics of the decentralization process itself may affect the prospects for successful implementation. Two necessary conditions of a decentralization initiative are devolution of authority and responsibility from the center; and an acceptance of that authority and responsibility by the local or regional units. Whether these both occur will depend in part upon why and how the decentralization takes place.

Top-down, bottom-up, or mutually desired devolution are ways of characterizing the decentralization initiative. In some cases, central government officials may have undertaken top-down resource management decentralization initiatives in order to solve their own problems – for example to reduce their political or financial responsibility in selected domestic policy areas (Simon 2002), or in response to pressure from external support agencies. In other cases, bottom-up pressure from stakeholders leads to the decentralization (Samad 2005). In still other cases, the decision to decentralize resource management to a lower and more appropriate level may have been the outcome of a process of mutual discussion and agreement between central officials hoping to improve policy outcomes and local stakeholders desiring greater autonomy or flexibility.

All other things being equal, it is anticipated that because decentralization initiatives require active basin-level stakeholder involvement, they are more likely to be implemented successfully if devolution is a mutually desired process involving both central government and local stakeholders.

Existing local-level governance arrangements contribute to continuation. The literature suggests that decentralization initiatives are more likely to be accompanied by active involvement, resource commitment, and compliance of basin stakeholders if existing community (village, tribe) governance institutions and practices are recognized and incorporated into the decentralization process. Also, transaction costs are likely to be lower if familiar organizational forms are used, rather than an additional set of organizational arrangements requiring greater adjustment on the part of stakeholders (though some new institutions will often need to be created to promote communication and integrate decisionmaking across communities within a river basin).

All other things being equal, decentralization initiatives are more likely to succeed in gaining stakeholder acceptance if they take advantage of existing social capital by being based upon, and constructed from, traditional community governance institutions and practices.

Central-Local Relationships and Capacities

Because successful decentralization requires complementary actions at the central government and local levels, other aspects of the central-local relationship can be expected to affect that success. Accordingly, a set of political and institutional variables were included relating to the respective capacities of the central government and the basin-level stakeholders, and with the relationship between them.

The extent of devolution of responsibilities and decisionmaking. A decentralization policy initiative announced by a central government may be only symbolic, while the central government retains in practice control over all significant resource management decisions; or it may represent an abandonment of central government responsibility for resource management without a concomitant establishment of local-level authority. In both cases stakeholder willingness to commit to decentralization and sustain involvement will be undermined. In better situations, the central government transfers degrees of both authority and responsibility for resource management to the stakeholders.

All other things being equal, greater prospects for success are expected where both authority and responsibility are devolved in tandem.

Financial autonomy and financial resources at the basin level reflect ability to implement decentralization. Decentralization of water resource management to the lowest appropriate level means that basin-level organizations have at least some autonomy to determine how funds shall be spent on resource management activities. On the other hand, decentralization does not have to mean that basin-level organizations and their members become solely responsible for all resource management funding. As already noted, one of the indicators of central government support for a decentralization policy can be its willingness to provide financial assistance to basin-level organizations without maintaining intrusive control over basin-level spending decisions. Therefore, while logic and experience suggest that basin-level organizations must have some degree of financial autonomy and some extent of financial resources in order for decentralization initiatives to be implemented successfully, the impact may not be linear.

All other things being equal, the prospects for success are expected to be hill shaped, with complete central government funding and control at one pole and complete basin-level funding and control at the opposite pole.

Local autonomy in institutional reform is the extent to which local communities can design and implement their own institutional arrangements, encouraging stakeholder involvement and reducing implementation costs through use of information better obtained at local level. However, as stakeholders create more institutional arrangements (particularly organizations and agencies), they incur greater transaction costs of maintaining all of them and coordinating their activities (Oechssler 1997).

All other things being equal, successful and sustainable implementation of decentralization initiatives is expected to occur more often in settings where local-level stakeholders are empowered to craft institutional arrangements for resource management at the basin and subbasin levels (including cross-jurisdictional arrangements), and modify them as needed.

Local-level experience with self-governance and service provision. In any country, the decentralization of water resource management does not occur in a vacuum. The ability of central government officials to strike a balance between supportiveness and intrusiveness, and the capacity of basin-level stakeholders to organize and sustain institutional arrangements, will in part be a function of their experiences with respect to other public services or responsibilities. The ability of central and local participants to perform successfully will depend on the skills and experiences they have developed.

It is expected that water resource management decentralization initiatives are more likely to be implemented successfully in settings where local participants have experience in governing and managing other resources or public services, for example land uses, schooling, and transportation.

Economic, political and social differences among basin users. In many countries, the distribution of political influence will be a function of economic, religious, or other social and cultural distinctions, which will also affect stakeholder communication, trust, and willingness to cooperate. These distinctions cannot be made to disappear but decentralization will be more successful where their adverse effects can be minimized.

All other things being equal, the greater and more contentious the economic, political, and social distinctions among basin users, the more difficult it will be to develop and sustain basin-scale institutional arrangements for governing and managing water resources.

Adequate time for implementation and adaptation. While it is obvious that longevity of water resource management arrangements may reflect their success, it is less obvious that their success may depend on their longevity. Time is needed to develop basin-scale institutional arrangements, to experiment with alternatives and engage in some trial-and-error learning, and to build trust. The relationship between time and success in water resource management is complicated; it is difficult to achieve a balance between the need for adaptability and innovation on the one hand, and the need for patience to allow sufficient time for reforms to consolidate on the other hand.

A curvilinear relationship may be expected, in which successful implementation is less likely to be observed among decentralization initiatives that are very young, is more likely at longer periods, but could taper off if central government and basin-level arrangements have proved insufficiently adaptable over long periods.

Basin-Level Institutional Arrangements

Successful implementation of decentralized water resource management may also depend on features of the basin-level arrangements created by stakeholders or by the central government.

Presence of basin-level governance institutions may be a prerequisite for successful water resource management. Sustained and effective participation of stakeholders presupposes the existence of arrangements by which stakeholders articulate their interests, share information, communicate and bargain, and take collective decisions. Basin-level governance is essential to the ability of water users to operate at multiple levels of action, which is a key to sustained successful resource preservation and efficient use (Ostrom 1990).

Because the existence of governance arrangements is a necessary, not sufficient, condition of successful resource management, success should not be expected everywhere there are basin-level governance institutions, but failure should be expected everywhere they are absent.

Recognition of subbasin communities of interest. The water management issues in a basin are viewed differently by the stakeholders that share the resource in various parts of the basin, according to their location and particular requirements. For example, downstream users' perspectives on water quality differ from those of upstreamers. Thus, while basin-level governance and management arrangements are essential to decentralized water resource management, the ability of subbasin stakeholders to address subbasin issues may be as important.

All other things being equal, successful implementation of basin decentralization is expected to have a positive relationship with level of participation of stakeholders in the process.

On the other hand, transaction costs may rise as such participation is institutionalized, as a large number of stakeholder organizations within the basin will bring greater coordination costs.

Here too, then, a hill-shaped relation of this variable to successful decentralization may be expected, with the absence of subbasin organizations and large numbers of subbasin organizations associated with lower success and greater prospects for success in between.

Information sharing and communication. The importance of information – more particularly, information symmetry – and opportunities for communication to the emergence and maintenance of cooperative decisionmaking is relatively well understood. In water resource management especially, where there can be so many indicators of water resource conditions and the performance of management efforts, forums for information sharing are vital to reducing information asymmetries and promoting cooperation.

All other things being equal, successful decentralized water resource management is expected to be more likely where information sharing and communication among stakeholders are more apparent.

Mechanisms for conflict resolution are needed to resolve disagreements between resource users about such issues as how well their interests are being represented and protected, the progress of the resource management program, and the distribution of benefits and costs. The success and sustainability of decentralized resource management efforts require the presence of forums for addressing such conflicts.

All other things being equal, successful implementation of decentralized water resource management is expected to be more likely in settings where forums for conflict resolution exist.

The set of variables and their hypothesized impact on the process and performance of decentralization in river basin management, as was developed in this section, will be inferred by applying several statistical tests to data collected from 83 river basins around the world. The data collection process and the construction of the workable variables are presented in the next sections.

3.3
Data and Variable Formation

A set of survey questions was developed corresponding to the list of variables presented in the previous section. These were compiled into a questionnaire, in English, Spanish, French, and Portuguese, which was sent to 197 river basin organizations around the world. A website with the capability of accommodating an online response was set up.[1] Questionnaires were returned by 103 respondents (52 percent response rate). After reviewing all responses[2] for completeness and accuracy of data 20 basins were removed, leaving 83 responses to comprise the final dataset. A tally of the survey responses is provided in Table 3.1.

3.3.1
Distributional Facts of the Sample River Basin Organizations

The 83 retained questionnaires reflect a representative distribution of river basin organizations both across continents and with regard to several basic characteristics

---

[1] *www.worldbank.org/riverbasinmanagement.*

[2] Transboundary river basin management processes are significantly different from institutional processes in national basins. For this reason, a decision was taken to not include transboundary basins in the analysis. Three basins that could be considered transboundary were included, however, because they were either mostly contained in one country or featured a river basin organization that had been developed only in one country.

Table 3.1. Distribution of responses and data collection efforts by continent

Continent	Questionnaires sent	Responses	Eliminated	Retained in dataset[a]
Africa and Middle East	18	14	2	12 (66)
Latin America	118	37	2	35 (30)
North America	5	5	0	5 (100)
East Asia-Pacific	7	7	3	4 (57)
Europe	49	40	13	27 (55)
Total	197	103	20	83 (42)

[a] Numbers in parentheses indicate percentage of retained questionnaires from the number that were sent.

in the sample. Latin America and Europe, which are the leading regions in reform of river basin management, account for 75 percent of the sample. Of the 83 basins, 32 are located in developed countries and 51 are located in developing countries. Basin size ranges from less than 10 000 square kilometers to, in three cases, over 1 million square kilometers. About two thirds of the basins have a population in the range 1 to 10 million. Mean year of creation of basin organizations in developed countries was 1979, and in developing countries 1989, with one organization created as early as 1927.

A preliminary analysis of basic correlations among key variables to discover whether there were some a priori relationships that required consideration found little significant difference between developed and developing countries as regards such matters as length of process of decentralization, type of governing body formed, and performance of the decentralization process. It was concluded, consistent with the analytical framework, that what actually affects the process and performances are the contextual and initial conditions on the one hand and the characteristics of central-local relationships and capacities on the other. Therefore, no distinction between developed and developing countries was included in the empirical analysis.

3.3.2
Data Manipulation

The questionnaire consisted of 47 questions producing 226 primary variables. They were divided into four groups: (*a*) general data variables that provide information on river basin organization contact information; (*b*) institutional setup variables that describe various aspects of the institutional arrangements in the basin before and after the decentralization process; (*c*) finance variables addressing aspects of the organization's budget; and (*d*) performance indicators that measure various performances of the organization. A detailed explanation and description of each variable is provided in Annex 3 of Dinar et al. 2005.

Some of the variables in the dataset are naturally correlated to each other. A principal component (PC) analysis was conducted in order to simplify the dataset by combining correlated variables, thus preventing a possible multicolinearity (inclusion of variables that correlate highly with other variables, leading to a biased estimate). A total of 15 PC variables were thus constructed.[3] Another type of manipulation of the primary variables was the creation of indexes to reflect values that are better expressed on a relative basis rather than an absolute scale.

Several questions were not answered by all river basin organizations, and thus several variables have a significant number of missing values and cannot be used. The primary and manipulated variables that were used in the analysis are presented in Annex 3 of Dinar et al. 2005, along with the hypotheses regarding their impact on the decentralization process.

3.4
Empirical Analysis

This section provides the general framework for the statistical analysis of the data collected from river basins around the world. The empirical approach used includes several procedures and sets of estimated equations applied in a regression analysis in order to understand the process and the performance of decentralization and the interaction between them. Appendix A3.2 to the present chapter contains a list of variables used and their designations; Table A3.7 in Appendix A3.1 gives descriptive statistics of the variables.

3.4.1
The Empirical Models

The main focus in the empirical analysis is on two types of relationships and their associated equations. The first, Eq. 3.1, explains a certain phenomenon in the basin, namely specifics of the decentralization process (*P*);[4] the second, Eq. 3.2, explains the

---

[3] For each PC analysis the following procedure was applied: Let $e_{i,k}^j$ be the value of the estimated i-th component of the j-th eigenvector of PC variable k. The value of variable k can then be calculated as

$$PC_k = \sum_{i=1}^{l_k} V_i \cdot \varepsilon_{i,k}^j \quad j=1,\ldots,J; \ \forall k$$

The estimated PC coefficients were used to create the PC variables, using the first eigenvector from the PC analysis. The eigenvectors of the first principal components, which were used in the creation of the principal component variables, explain between 25 and 99 percent of the standardized variance among the variables.

[4] Equation 3.1:

$$P = g(C, R, I \mid X) \tag{3.1}$$

where
P is a vector of characteristics of the decentralization process
C is a vector of contextual factors and initial conditions
R is a vector of characteristics of central-local relationships and capacities
I is a vector of internal configuration of basin-level institutional arrangements
X is a vector of "other" variables, identified as necessary.

level of success or progress of the decentralization process (S).[5] The two relationships are linked in that P is a dependent variable in the first equation and an independent variable in the second equation, allowing the formulation of an equation system linking P and S (Eq. 3.3).[6]

Various statistical methods may be used to measure the degree of success, depending on the nature of the variable S. Based on the discussion in previous sections, one possible way is to use a dichotomous variable that takes the value 1 when decentralization was initiated and 0 when no decentralization took place in spite of government intent.

A second way of measuring success is to quantify normatively the extent of achieving several important original goals of the decentralization process by rank-ing the various basin organizations on a scale ranging from $\underline{s}$ to $\bar{s}$ in terms of the decentralization success, which allows S to have more than the two dichotomous values.

A third way of measuring progress of decentralization is by comparing performance of certain decentralization objectives between the present and the predecentralization period. Performance variables may include level of participation, local responsibility, financial performance, and economic activity. Using this definition, S becomes a continuous variable.

3.4.2
Empirical Specifications of the Decentralization Process and Its Performance

As regards the specification of Eq. 3.1 explaining the characteristics of the decentralization process, some variables are of special interest as they contrast observations across river basin decentralization processes under a variety of situations: they include length of the decentralization process, transaction costs of the processes, and level of involvement of stakeholders. During this analysis various statistical procedures were used, according to the nature of the variable being analyzed.[7]

Several variables were identified as serving to measure decentralization success or progress in Eq. 3.2: these included whether or not institutional change occurred; success in achieving such major objectives as conflict resolution, flood control, and water scarcity improvement; and incremental change, before and after decentralization,

---

[5] Equation 3.2:

$$S = f(C, P, R, I \mid X) \tag{3.2}$$

where S is a measure of success/progress of the performance of the decentralization of management in the river basin.

[6] Equation system (Eq. 3.3), with all parameters having the same meaning as in Eqs. 3.1 and 3.2:

$$\left\{ \begin{array}{l} P = e(C, R, I \mid X) \\ S = h(C, P, R, I \mid X) \end{array} \right. \tag{3.3}$$

[7] For example, TOBIT or OLS procedures were used for length of decentralization process, as values of that variable are continuous; and a GLM procedure was applied to string variables related to creation or dismantling of institutions during the decentralization process.

in the way in which various tasks and responsibilities were carried out. Again, statistical procedures appropriate to the three different methods of measuring success or progress described in the previous section were used, according to the nature of the variable being analyzed.[8]

3.5
Results of the Statistical Analysis

The results of the statistical analysis are presented in this section. They are split into three subsections: the equations describing the decentralization process, the equations describing the decentralization performance, and the equation systems describing the interaction between the decentralization process and the level of performance.

3.5.1
Decentralization Process

Several relationships were identified that could shed light on the characteristics of the decentralization process in terms of its length, complexity, and participation of stakeholders. The hypothesized relationships between the variables in the empirical analysis, following Eq. 3.1, are based on the discussion in Sect. 3.2. Five equations were estimated representing various aspects of the decentralization process: institutions created, institutions dismantled, involvement of water groups, political cost of the process, and length of the process (Table A3.1, Appendix A3.1 to the chapter). The explanatory variables used in the analysis are generally in line with the hypothesized signs; certain variables included in more than two equations, such as that related to the percentage of users paying their tariffs, are significant, and their signs are as expected.[9]

3.5.2
Decentralization Performance

The estimated decentralization performance equations, following Eq. 3.2, are presented in three tables, using different estimation procedures and explanatory and dependent variables (Tables A3.2, A3.3, and A3.4, Appendix A3.1). Generally speaking, the various estimated model equations display a robust set of results that support the hypotheses. Only one coefficient – that related to the percentage of the basin's budget obtained from basin stakeholders – displayed the opposite sign to that expected. Results indicate a significant positive correlation between decentralization success and share of the budget spent in the basin; share of users paying their tariffs; comprehensiveness of decentralization reform objectives; and water scarcity level in

---

[8] For example, a LOGIT procedure was applied to the dichotomous variable measuring whether or not an institutional change had occurred; a TOBIT procedure was applied to the variables measuring degree of success in achieving various objectives; and an OLS procedure was applied to the "before and after decentralization" variables.

[9] Given the nature of the five dependent variables, different signs of the same explanatory variable are expected, depending on the equation.

the basin.[10] Top-down initiation of decentralization reform was found to be detrimental to the success of the decentralization process.

3.5.3
Interaction between Decentralization Process and Its Level of Performance

In order to assess the impact of the decentralization process on the level of success, several equation systems consisting of two equations were estimated, following Eq. 3.3. The first equation depicted the decentralization process, as was described in the previous section, and then the dependent process variable was included as an instrumental variable in a relationship that described the performance of the decentralization process (Tables A3.5 and A3.6, Appendix A3.1). The statistical results follow more or less the lines that were reported in the case of the single equation models. However, several of the important coefficients were not significant in the system equation compared with the single equation estimates. The meaning of these results will be discussed in the following section.

3.5.4
Interpretation of Results

The results of the analysis are grouped according to the four sets of variables that formed the basis for the hypotheses developed in Sect. 3.2 above.

Contextual Factors and Initial Conditions

Variables in this group significantly affect the nature of the decentralization process and its performance. It is suggested that:

- The greater the extent of initial decentralization in the basin, the less time the decentralization process took.
- The greater the number of major problems in the basin prior to decentralization, the greater the extent of reported improvement between before and after decentralization.
- The greater the reliance on surface water in the basin, the higher the degree of water user involvement and participation; the larger the number of institutions that were created during the decentralization process; the greater the political transaction costs associated with the decentralization process; and the greater the extent of reported improvement between before and after decentralization.
- The greater the water scarcity problem in the basin, the less time decentralization took; the greater the extent of reported improvement between before and after decentralization; and the greater the extent of reported success with respect to the major objectives of basin management. The robustness of the water scarcity variable is of particular interest as it fits nicely with the notion of "scarperation"

---

[10] Notice that the lower the ratio of rainfall to evaporation, the higher the scarcity, so a negative sign of the scarcity coefficient in the equation reads as a higher level of scarcity.

(Dinar and Dinar 2005), which suggests that scarcity is an incentive for cooperation among the parties involved.

Rich and well-endowed basins do not necessarily have an advantage over less-endowed basins; stressed resource conditions and the presence of multiple major problems can be stimulants to effective action. Case-specific approaches may lead to similar performances despite very different starting points.

Characteristics of the Decentralization Process

Variables in this group, such as the political economy of the process, participation, compliance, and governance level, suggest an array of supporting results, as follows:

- The greater the extent of tariff compliance the lower the political transaction costs of the decentralization process; and the smaller the number of institutions that were dismantled during the decentralization process.
- The larger the number of water use sectors present in the basin, the larger the number of institutions that were created during the decentralization process.
- The greater the availability of forums for dispute resolution, the greater the extent of water user involvement and participation.
- The larger the number of types of disputes in the basin, the greater the extent of reported improvement between before and after decentralization.
- The greater the political transaction costs associated with the decentralization process, the smaller the reported improvement between before and after decentralization; and the less likelihood that some form of institutional change was associated with the decentralization process.
- The greater the number of institutions dismantled during the decentralization process, the greater the reported improvement between before and after decentralization.
- The longer the decentralization process took, the greater the extent of reported improvement between before and after decentralization.
- The more comprehensive the basin management objectives were, the greater the extent of reported success with respect to the major objectives of basin management; and the greater the reported improvement between before and after decentralization.

These results indicate that diverse and crowded basins do not necessarily have to face higher political cost and lower levels of performance of the reform, if an appropriate set of mechanisms and objectives, such as forums for dispute resolution, and a coherent reform agenda, are put in place at the appropriate time and over an appropriate time period (which may be long rather than short).

Central-Local Relationships and Capacities

Variables included in this set, such as budget and funding by the government agencies and the initiation of the reform process, are also consistent with expectations. The common findings include:

- The larger the share of the basin organization budget received from external governmental agencies, the smaller the number of institutions that were created during the decentralization process.
- The greater the share of the organization budget coming from external governmental agencies, the greater the reported improvement between before and after decentralization.
- The more top down the decentralization process was, the smaller the extent of reported success with respect to the major objectives of basin management.

This group of findings indicates that decentralization is more likely to be successful where central government allows stakeholders to initiate and lead the reform process, with a certain amount of budget support.

Basin-Level Institutional Arrangements

This set of variables includes local configurations such as user group presence and budget sources and usage. Common finding across the regression models suggest that:

- The greater the presence of existing user groups in the basin, the greater the reported improvement between before and after decentralization.
- The greater the share of the basin organization budget spent within or returned to the basin, the greater the extent of reported improvement between before and after decentralization; the more likely that some form of institutional change was associated with the decentralization process; and the greater the extent of reported success with respect to the major objectives of basin management.
- The greater the share of the basin organization budget contributed by other sources, the smaller the extent of water user involvement and participation.
- The larger the share of the budget collected from basin stakeholders, the greater the number of institutions that were dismantled during the decentralization process; and the longer the decentralization process took.
- The greater the budget per capita, the lower the extent of reported success with respect to the major objectives of basin management.

The results of this group of variables indicate that presence of water user organizations can positively impact the decentralization process, even if their involvement makes the process longer. Also, the basin organization budget is an important tool for management and for encouragement of participation, and can significantly enhance the decentralization process if well designed and managed. Again, success in this respect is not confined to well-endowed basins.

3.6
Conclusion and Policy Implications

The statistical estimates suggest that both the process of decentralization and its performance level, as measured by several variables, are well explained by a set of explanatory variables. Several independent variables provide a robust explanation regardless of the equation selected and the estimation procedure used, and merit further attention.

Water scarcity is an important variable that affects the process as well as the performance of decentralization. As water in the basin is less abundant incentives for a simpler decentralization process and a more successful outcome are more likely. The presence of scarcity may therefore be a stimulus to reform, uniting the stakeholders in the basin.

In addition to water scarcity, the number and severity of other water resource problems present in a basin prior to decentralization was (perhaps surprisingly) a positive factor with respect to both the initiation of decentralization reforms and their perceived success. The more ambitious and nearly comprehensive the decentralization effort was, and the greater the problems users faced, the more likely they were to see the effort as worthwhile and effective.

Existence and number of organized user groups was positively associated with the initiation of decentralization reforms, but also with the costs and difficulty of achieving decentralization. Existence of dispute resolution mechanisms was positively associated with water user involvement and with perceived decentralization performance. Length of the decentralization process was positively associated with perceptions of decentralization success and with tariff compliance and share of the basin organization budget contributed by stakeholders. A decentralization process that was characterized by protracted political struggle leaves a negative impact on the decentralization performance.

Dismantling of institutions during the decentralization process contributes to the performance of the decentralization process. Combined with the two preceding findings, it appears that complexity and conflict are two distinct characteristics and work in opposite ways. The mere presence of a larger number of organizations within a river basin, and the sheer length of time a decentralization reform takes, do not appear to be substantial negative factors. On the other hand, highly conflictual decentralization processes are associated with poorer performance, and some elimination of previously existing institutional arrangements may be a positive factor. Thus what matters is not so much how complicated or lengthy the process is, but the degree of conflict and the ability to make organizational changes along the way.

River basins with higher percentages of their budgets from external governmental sources (such as the local and federal governments) benefit from better stability and support and it shows in the performance of the decentralization process, although the same relationship does not hold for the budget share contributed by other outside sources.

In basins where stakeholders accepted greater financial responsibility, complying with tariffs and contributing to the budget for basin management, decentralization process and performance measures increased. Combined with the preceding finding, it appears that the financial dimensions of decentralized river basin management are both important and complex: it is the combination of financial responsibility (on the part of water users), financial autonomy (basin revenues remaining in the basin), and central government support that is associated with success, and not necessarily one element alone. This is consistent with the analytical framework, which hypothesized that a configuration of factors that included a supportive but not controlling role for the central government, and responsibility but not complete independence for the water users in the basin, would be associated with successful implementation of decentralization reforms.

Appendix A3.1 – Statistical Tables

In Tables A3.1–A3.6, values in parentheses are t-values: $t > 2.326$ indicates significance at 1 percent; $2.325 > t > 1.646$ indicates significance at 5 percent; $1.644 > t > 1.282$ indicates significance at 10 percent.

Table A3.1. Equations describing some features of the decentralization process

Estimation procedure	GLM	OLS	GLM	POISSON	GLM
Dependent variable	InstCreatd	WuasInvlv	PltclCost	YrsDecentral	InstDismntld
Independent variable					
Intercept	0.461 (0.76)	0.045 (0.41)	1.183 (1.97)	1.745 (1.96)	0.321 (1.99)
%BgtBsn					1.456 (7.08)
%BgtExtr	−1.001 (−3.45)				
%BgtSpnt		−0.093 (−1.31)			−0.497 (4.45)
%BgtSrcs		−0.212 (−2.66)			
%UsrPay			−0.007 (−3.35)	0.017 (4.58)	−0.012 (−8.91)
Facilities			0.001 (2.57)		
FormsDisput1		0.068 (2.53)			
FormsDisput2	0.319 (1.53)				
GovrBdy	−0.052 (−0.81)	−0.028 (−1.62)		−0.498 (−3.48)	
MainObj				0.510 (1.05)	
MinrObj	0.016 (1.44)		−0.011 (−1.31)		
PrblmsBfr	−0.056 (−0.79)		−0.131 (−1.63)		
Scarcity1				−0.009 (−3.79)	
SectrComposit	0.519 (3.66)	−0.006 (−0.16)			0.179 (2.04)
ShareSW	1.864 (6.41)	0.395 (4.61)	0.482 (1.88)		
TypesDisput					−0.063 (−2.76)
WuasInvlv	−0.126 (−0.36)		0.259 (0.80)		

Table A3.1. *Continued*

Estimation procedure	GLM	OLS	GLM	POISSON	GLM
Dependent variable	InstCreatd	WuasInvlv	PltclCost	YrsDecentral	InstDismntld
Independent variable					
Log Pseudolikelihood	−115.59		−105.84	−354.38	−73.37
F-test		11.01			
Adjusted R^2		0.423			
Wald chi-square				28.42	
Pseudo-R^2				0.241	

Table A3.2. GLM and TOBIT equations of the decentralization performance

Estimation procedure	GLM	TOBIT	GLM	TOBIT
Dependent variable	SuccObj1	SuccObj1	SuccObj2	SuccObj2
Independent variable				
Intercept	−0.305 (−0.12)	13.183 (1.52)	−0.802 (−2.00)	−0.900 (−3.34)
%BgtBsn		−13.335 (−1.96)	1.077 (2.46)	0.398 (2.21)
%BgtExtr	8.744 (3.50)	15.063 (2.28)	0.726 (1.35)	
%BgtSrcs		−19.373 (−2.04)		
%UsrPay			0.019 (1.67)	0.01 (2.84)
BgtPrCpta			−0.155 (−2.56)	
Facilities		0.0004 (1.35)		
FormsDisput1	1.446 (2.10)	0.272 (4.24)		
FormsDisput2		0.323 (2.24)		
InstDismntld		−0.460 (−3.87)		
MainObj	2.809 (2.28)	1.099 (6.67)	3.085 (12.55)	1.451 (9.36)
MinrObj	0.147 (1.59)	0.042 (6.08)	0.027 (1.51)	0.015 (2.12)
MtdCreatn		−0.335 (−3.54)		−0.189 (−2.36)
PltclCost				0.116 (1.50)
Scarcity1	−0.019 (−1.80)	−0.009 (−1.46)	−0.020 (−3.63)	−0.014 (−1.72)
SectrComposit	−2.426 (−2.34)	−0.297 (−3.22)		
SectrUseShars		0.199 (2.15)		
ShareSW	−4.055 (−1.82)	−0.424 (−2.35)		
TypesDisput		−0.129 (−4.43)		
YrCreation		−0.006 (−1.55)		

Table A3.2. *Continued*

Estimation procedure	GLM	TOBIT	GLM	TOBIT
Dependent variable	SuccObj1	SuccObj1	SuccObj2	SuccObj2
Independent variable				
YrsDecentral		−0.022 (−2.12)		
LR chi-square		392.76		159.47
R^2 or pseudo R^2		0.593		0.385
Log likelihood	−271.69	−167.95	−139.70	−127.27

Table A3.3. OLS equations of the decentralization performance

Estimation procedure	OLS							
Dependent variable	PrblmsAftr	PrblmsAftr	ImprvRespons	ImprvRespons	IncrmntTasks	IncrmntTasks	IncrmntImprv	IncrmntImprv
Independent variable								
Intercept	1.942 (2.94)	1.950 (2.95)	0.001 (0.00)	0.056 (0.06)	0.062 (0.86)	0.066 (1.28)	0.198 (1.81)	0.191 (1.86)
%BgtExtr							0.255 (1.56)	0.255 (1.11)
%BgtSpnt	0.536 (1.80)	0.471 (1.67)	2.134 (2.64)	2.100 (2.78)				
%BgtSrcs	−0.483 (−1.53)	−0.496 (−1.25)						
ExistUsrGrp	0.432 (2.44)	0.388 (2.04)						
GovrBdy	−0.091 (−1.14)	−0.085 (−1.00)						
InstDismntld					0.116 (2.66)	0.119 (3.29)	0.161 (2.47)	0.165 (1.94)
MainObj	0.300 (1.60)	0.312 (1.66)	1.357 (2.47)	1.376 (2.47)				
PltclCost			−0.789 (−1.94)	−0.817 (−1.84)	−0.170 (−4.37)	−0.172 (−1.93)	−0.240 (−4.23)	−0.240 (−1.55)
PrblmsBfr	0.470 (6.73)	0.477 (5.82)						
Scarcity1		−0.001 (−0.68)		−0.104 (−2.43)		−0.001 (−2.50)		−0.001 (−3.01)
SectrComposit							−0.098 (−1.40)	−0.093 (−0.77)
ShareSW			1.640 (1.60)	1.721 (1.59)				
TypesDisput			−0.197 (−1.54)	−0.204 (−1.46)	0.018 (1.48)	0.018 (1.88)		
YrsDecentral	0.028 (1.56)	0.026 (2.39)						
F-test	10.43	12.35	4.1	4.46	8.93	3.10	5.83	2.69
Adjusted R^2	0.449	0.485	0.162	0.220	0.227	0.258	0.192	0.232

Table A3.4. LOGIT equations of the decentralization performance

Estimation procedure	LOGIT	LOGIT
Dependent variable	InstCng	InstCng
Independent variable		
Intercept	−1.375	−1.370
	(−2.99)	(−1.27)
%BgtSpnt	1.400	1.462
	(1.91)	(2.22)
InstDismntld	1.198	1.266
	(2.99)	(4.10)
PltclCost	−0.753	−0.762
	(−2.52)	(−2.35)
PrblmsBfr	0.198	0.191
	(1.32)	(1.30)
Scarcity1		−0.005
		(−0.86)
LR/Wald chi-square	23.94	22.71
Pseudo R^2	0.222	0.225

Table A3.5. Results of a manually estimated GLM-Poisson equation system of decentralization process-performance

Decentralization process equation[a]				Decentralization performance equation[a]			
Variable	(a) WuasInvlv	(b) YrsDecentral	(c) InstDismntld	Variable	(A) SuccObj1	(B) SuccObj1	(C) SuccObj1
Constant	0.045 (0.41)	1.745 (1.96)	0.321 (1.99)	Constant	7.707 (0.10)	18.99 (0.29)	20.854 (0.35)
%BgtBsn		1.456 (7.08)		%BgtBsn	−79.490 (−1.22)	−82.04 (−1.23)	−89.531 (−1.28)
%BgtExtr				%BgtExtr	92.165 (1.33)	93.97 (1.32)	101.08 (1.37)
%BgtSpnt	−0.093 (−1.70)		−0.497 (−4.45)	%BgtSrcs	−116.56 (−1.25)	−120.19 (−1.26)	−130.82 (−1.31)
%BgtSrcs	−0.212 (−2.58)			InstDismntld	−1.928 (−1.26)	−1.842 (−1.38)	−1.995 (−1.12)
%UsrPay		0.017 (4.58)	−0.012 (−8.91)	MainObj	3.116 (2.16)	3.121 (2.23)	2.958 (2.16)
FormsDisput1	0.068 (2.60)			Scarcity1	−0.020 (−1.77)	−0.020 (−2.44)	−0.021 (−2.45)
GovrBdy	−0.028 (−1.56)	−0.498 (−3.48)		SectrComposit	−1.932 (−1.67)	−1.963 (−1.66)	−1.927 (−1.38)
MainObj		0.510 (1.05)		ShareSW	−4.985 (−0.66)	−4.328 (−1.11)	−4.246 (−1.03)
Scarcity1		−0.009 (−3.79)		TypesDisput	−0.533 (−1.58)	−0.497 (−1.59)	−0.526 (−1.72)
SectrComposit	−0.006 (−0.15)		0.178 (2.04)	YrsDecentral	−0.097 (−1.28)	−0.136 (−0.73)	−0.078 (−1.32)
ShareSW	0.395 (4.21)			WuasInvolv	1.578 (0.15)		
TypesDisput			−0.063 (−2.76)	MinrObj	0.193 (1.83)	0.194 (1.87)	0.193 (1.80)
YrsDecentral				MtdCreatn	−1.276 (−1.56)	−1.320 (−1.53)	−1.259 (−1.58)
				FormsDisput1	1.851 (1.55)	2.018 (1.41)	1.744 (1.44)
				SectrUseShars	0.448 (1.09)	0.304 (0.74)	0.232 (0.59)
				FormsDisput2	1.331 (0.82)	1.126 (0.74)	0.870 (0.62)
				YrCreation	−0.003 (−0.10)	−0.009 (−0.29)	−0.009 (−0.33)

Table A3.5. *Continued*

Decentralization process equation[a]				Decentralization performance equation[a]			
Variable	(a) WuasInvlv	(b) YrsDecentral	(c) InstDismntld	Variable	(A) SuccObj1	(B) SuccObj1	(C) SuccObj1
				Facilities	0.001 (0.42)	0.001 (0.54)	0.001 (0.59)
Estimation procedure	GLM	Poisson	GLM	Estimation procedure	GLM	GLM	GLM
Log Pseudo-likelihood	−6.916		−73.373		−267.49	−267.81	−268.49
Pseudo R^2		0.241					
Wald chi^2		28.42					

[a] The equation systems that were estimated consist of equations aA, bB, and cC.

Table A3.6. Results of a simultaneously estimated OLS equation system of decentralization process-performance

	Equation 1		Equation 2	
	WuasInvlv	PrblmsAftr	WuasInvlv	ImprvRespons
Constant	−0.033 (−0.19)	1.945 (2.89)	0.019 (0.15)	−0.913 (−1.06)
%BgtSpnt	−0.09 (−1.20)	0.404 (1.24)	−0.100 (−1.42)	2.293 (2.71)
%BgtSrcs	−0.219 (−2.70)	−0.616 (−1.58)	−0.234 (−2.94)	
ExistUsrGrp	−0.001 (−0.02)	0.415 (2.14)		
FormsDisput1	0.076 (2.64)		0.081 (2.91)	
GovrBdy	−0.024 (−1.27)	−0.087 (−1.06)	−0.025 (−1.34)	
MainObj	−0.046 (−0.95)	0.316 (1.65)	−0.038 (−0.81)	1.286 (2.34)
PltclCst			0.058 (1.75)	−0.895 (−2.20)
PrblmsBfr	0.014 (0.84)	0.489 (6.61)		
SectrComposit	−0.018 (−0.44)		−0.01 (−0.27)	
ShareSW	0.383 (4.16)		0.360 (4.15)	1.304 (0.90)
WuasInvlv		−0.427 (−0.58)		0.902 (0.39)
YrsDecentral	0.04 (0.99)	0.029 (1.53)	0.005 (1.26)	0.024 (0.49)
System F-test	8.75		2.81	
System adjusted-R^2	0.429		0.167	

Table A3.7. Descriptive statistics of the variables included in the working dataset

Variable	Mean	Standard deviation	Minimum	Maximum
%BgtBsn	0.356	0.391	0.000	1.000
%BgtExtr	0.190	0.328	0.000	1.000
%BgtSpnt	0.287	0.474	−0.267	1.370
%BgtSrcs	−0.115	0.413	−0.713	0.710
%UsrPay	2.084	13.990	0.000	127.886
BgtPrCpta	0.292	2.658	0.000	24.213
ExistUsrGrp	0.871	0.783	0.000	1.731
Facilities	21.762	263.256	−1 410.219	1 348.949
FormsDisput1	2.241	1.265	0.000	4.000
FormsDisput2	0.940	0.502	0.000	2.000
GovrBdy	3.566	1.768	0.000	5.000
ImprvRespons	1.466	3.639	−8.294	9.217
IncrmntImprv	0.104	0.518	−4.162	0.728
IncrmntTasks	0.128	0.371	−2.671	0.762
InstCng	0.627	0.487	0.000	1.000
InstCreatd	1.542	1.262	0.000	3.000
InstDismntld	0.627	0.837	0.000	3.000
MainObj	1.008	0.692	0.000	1.731
MinrObj	6.208	8.791	0.000	33.972
MtdCreatn	1.193	0.903	0.000	2.000
PltclCost	0.494	0.942	0.000	5.000
PrblmsAftr	5.837	1.524	2.334	8.967
PrblmsBfr	6.871	1.823	0.000	9.750
Scarcity1	5.056	27.965	0.000	246.679
Scarcity2	177.383	1 142.484	0.000	10 438.417
SectrComposit	1.257	0.793	0.000	2.183
SectrUseShars	0.299	0.768	−0.425	6.609
ShareSW	0.434	0.381	0.000	1.000
SuccObj1	3.540	7.814	0.000	63.510
SuccObj2	2.907	2.551	0.000	7.555
TypesDisput	4.207	2.934	0.000	9.000
YrCreation	1 985.229	18.486	1 926.000	2 002.000
YrsDecentral	2.711	7.547	0.000	36.000

Appendix A3.2 – Variables Used in the Analysis

%BgtBsn. Share of the basin's budget obtained from basin stakeholders.

%BgtExtr. Share of the basin's budget allocated by external (government) agency.

%BgtSpnt. Share of the budget spent in the basin and not returned to external government.

%BgtSrcs. Share of budget from sources other than government and basin stakeholders.

%UsrPay. PC variable measuring percentage of users in the irrigation, industrial, and urban sectors that pay their tariffs.

BgtPrCpta. Budget per capita in the basin.

ExistUsrGrp. PC variable measuring existence of irrigation, industrial, and domestic user groups (measured as individually dichotomous variables) in the basin.

Facilities. PC variable incorporating the values (length or quantity and capacity) of canals, reservoirs, dams, and treatment facilities into one "facilities" variable.

FormsDisput1. Forums available to hear disputes.

FormsDisput2. Measures, using a dichotomy variable, the existence of dispute resolution institutions.

GovrBdy. Distinguishes between levels of governance of the basin organization using values ranging 1–5 with increasing centralization.

ImprvRespons. PC variable taking into account the difference between after and before decentralization regarding five responsibilities, with higher values indicating greater success.

IncrmntImprv. PC variable measuring incremental improvement in various problems in the basin before and after decentralization.

IncrmntTasks. PC variable measuring incremental change in eight variables measuring tasks of local and basin-level management before and after decentralization.

InstCng. Dichotomous variable measuring whether or not there was an institutional change associated with the decentralization process.

InstCreatd. New institutions created in decentralization process.

InstDismntld. Institutions dismantled in decentralization process.

MainObj. PC variable comprising three main objectives of the basin organization related to conflict resolution, flood control, and water scarcity improvement.

MinrObj. PC variable comprising three minor objectives of the basin organization, a combination of a set of 25 possible minor objectives.

MtdCreatn. Variable indicating whether basin organization creation was bottom up or top down.

PltclCost. Variable measuring the political and transaction costs of the decentralization process via the creation of new institutions.

PrblmsAftr. PC variable measuring composite success of decentralization.

PrblmsBfr. PC variable measuring composite level of problems in several domains related to management issues in the basin: flooding, water scarcity, environmental quality, water conflicts, land degradation, and development issues.

Scarcity1. Variable reflecting one measure of scarcity, measured as the ratio between rainfall and evapotranspiration.

Scarcity2. Variable measuring available water resources per person residing in the basin.

SectrComposit. PC variable measuring composition of the subsectors in the basin: irrigation, industry, domestic, hydropower, and environment.

SectrUseShars. PC variable taking into account distribution of water use shares of the five main water-using sectors: irrigation, industry, domestic, hydropower, and environment.

ShareSW. Variable measuring share of surface water in the available water resources in the basin.

SuccObj1. PC variable capturing the integrated level of success of the three main objectives and the other 25 minor objectives.

SuccObj2. PC variable measuring the success of the three main objectives.

TypesDisput. Main types of disputes or issues that usually need resolving.

WuasInvlv. Variable assessing the degree of water user association involvement and participation.

YrCreation. Variable measuring the year in which the basin organization was created.

YrsDecentral. Variable measuring the length of the decentralization process.

Part II

The Case Studies

AUSTRALIA
MURRAY-DARLING RIVER BASIN

o   SELECTED CITIES
⊛   NATIONAL CAPITAL
——   MAJOR ROADS

This map was produced by the Map Design Unit of The World Bank.
The boundaries, colors, denominations and any other information
shown on this map do not imply, on the part of The World Bank
Group, any judgment on the legal status of any territory, or any
endorsement or acceptance of such boundaries.

AUSTRALIA

Tasman Sea

Murray-Darling
River Basin

AUSTRALIA

Gold Coast
Grafton
Brisbane
Port Macquarie
Newcastle
Richmond
Sydney
Nowra
CANBERRA
Goulburn
Bathurst
Orange
Parkes
Dubbo
Walgett
Gwydir
Namoi
Macquarie
Darling
Bourke
Wilcannia
Broken Hill
Thargomindah
Murrumbidgee
Wagga Wagga
Albury
Wodonga
Wangaratta
Bairnsdale
Sale
Traralgon
Moe
Marwell
Melbourne
Echuca
Bendigo
Melton
Ballarat
Geelong
Colac
Warrnambool
Portland
Mount Gambier
Horsham
Mildura
Darling
Murray
Loddon
Murray Bridge
Adelaide
Murray
Port Augusta
Leigh Creek

0   50   100 Kilometers
0   50   100 Miles

Chapter 4

Australia: Murray-Darling Basin

W. Blomquist · B. Haisman · A. Dinar · A. Bhat

4.1
Background

4.1.1
Introduction

Water resources are a major public issue in Australia because of their scarcity and extreme variability. Although the coastal fringes are relatively well endowed with water, and are therefore where most of the population resides, the interior is arid and water is very scarce, making Australia the driest inhabited continent on Earth. The Murray-Darling basin is an interior basin of southeast Australia, taking its name from the two dominant rivers, the Murray and the Darling. It is defined by the catchment areas of these rivers and their many tributaries.

In light of the high degree of development of water use in the basin, the dominant basin management issues of the 20th century were water scarcity, overallocation of water rights,[1] and drought exposure. These issues stimulated the development of certain institutional arrangements in the basin from the beginning of the 20th century to the 1990s. Those institutions provided for the management of water distribution through the issuance of water use licenses; the allocation of Murray River flows among the states of South Australia, New South Wales, and Victoria; the construction and operation of water storage facilities to conserve and regulate river flows; and in the latter decades of the century, moratoriums on the issuance of water licenses, and ultimately a cap on diversions from the Murray-Darling system.

The gradual decline in the health of the Murray-Darling system over this period indicated that a more innovative approach was required, and the integrated water resource management approach initiatives that have been introduced in response have had some degree of success in finding a balance between complex hydrological and institutional issues.

4.1.2
Basin Characteristics

The Murray-Darling basin lies to the west of the Great Dividing Range, which runs the length of the east coast of Australia. The basin extends across much of south-

---

[1] Almost half of the basin's surface water management areas are overappropriated; that is, authorized uses exceed mean annual flows.

eastern Australia, with the mouth of the Murray River on the southern coast of Australia near Adelaide. It includes over 1 million square kilometers, and about one seventh of the land area of Australia. The basin extends to over three quarters of the state of New South Wales, more than half of the state of Victoria, significant portions of the states of Queensland and South Australia, and includes the whole of the Australian Capital Territory. Well over half of the basin is in New South Wales and almost a quarter is in Queensland. The basin contains more than 20 major rivers as well as important groundwater systems. It is also an important source of freshwater for domestic consumption, agricultural production, and industry.

The rivers of the Murray-Darling basin are characterized by flat gradients (much of the basin is less than 200 meters above sea level), highly variable flows, and limited runoff. Average annual runoff is some 24 million cubic meters, of which around half is lost to evaporation and percolation. Total runoff is the lowest of any of the world's major basins and average annual flow to the sea is a mere 400 cubic meters per second. Much of the basin is semiarid and some 86 percent of the area contributes no runoff. The basin covers 14 percent of Australia but receives only 6.1 percent of Australia's mean annual runoff (Goss 2003).

Wetlands play an important role in the basin's rich ecosystem. There are about 30 000 wetlands in the basin, with 11 being listed for their internationally significant environmental values. The wetlands are major considerations in environmental management of the rivers.

The water resources of the basin are now highly developed. Annual diversions from the river system are 11.43 million cubic meters, 96 percent of which is for irrigation (Goss 2003). Total water storage capacity in the basin is 34.7 million cubic meters, which supports some 1 470 000 hectares of irrigated crops and pastures (representing 71 percent of Australia's total area of irrigated crops and pastures).

In 1996 the basin was home to nearly 2 million people (or about 11 percent of the total Australian population) and another million people outside the basin were heavily dependent upon the basin's water resources. The basin boasts a gross value of production of over A$23 billion, of which approximately A$4.5 billion is generated by irrigated agriculture.[2] Around 40 percent of Australia's gross value of agricultural production originates from the Murray-Darling basin.

4.1.3
Water Resource Problems

Many of the water resource problems in the Murray-Darling basin emanate from its predominantly semiarid nature. As described in the previous section, runoff is limited, and it is becoming increasingly difficult to find the supply to meet the growing demand in this economically active region.

A second (though related) major basin management issue is salinity. For climatological and geological reasons, much of Australia's soils store quantities of salt. In many irrigated areas, the application of water to the soil has elevated the concentration of salts in the underground water table and in surface water channels draining irrigated

---

[2] 10 Australian dollars = 7.4 US dollars (April 2006).

land. Compared with this irrigation-related salinity, however, a much larger and more difficult challenge is dryland salinity. A European model of land clearing for agricultural development was followed in Australia in the late 19th and 20th centuries. It has been estimated that 15 billion trees, which used to transpire the saline water their roots drew from the aquifers, were removed from the Murray-Darling basin alone. Saline groundwater levels in these areas have risen closer to the surface, sterilizing productive land in some places and boosting river salinities through surface runoff pathways. Currently it is estimated that 7 million hectares of Australia are affected by dryland salinity, and the National Land and Water Resources Audit projects that without intervention this will rise to around 17 million hectares by the year 2050.

There are other aspects to water quality deterioration in the basin. Awareness of the impaired status of aquatic and riparian species and habitat in much of the basin has risen since the 1970s and reached a level of concern at the outset of the 21st century that nearly matches that for water scarcity. In response to growing evidence of a decline in river health, the Murray-Darling Basin Commission sponsored work to bring together current studies and knowledge and to inform the debate on restoration of river health. The results (Norris et al. 2001) showed, among other things, that:

- 38 percent of the river length assessed had biota that was significantly impaired.
- 10 percent of the river length was found to be severely impaired, having lost at least 50 percent of the types of aquatic invertebrates expected to occur there.
- over 95 percent of the river length assessed in the Murray-Darling basin had an environmental condition that was degraded and 30 percent was considerably modified from the original condition.

This has set the scene for what the Basin Commission has termed the Living Murray initiative, which is a concerted attack on the failing health of the Murray River. The states were already addressing environmental flows in other streams in the basin, but were slower to address the Murray itself because its highly regulated nature as a transboundary river limits the options for modifying flows. Large reductions of water diversions for consumptive uses are expected to be necessary in order to achieve the levels of ecological restoration envisioned by the Living Murray initiative. Currently, there is a major focus of debate in Australia concerning the need to recover a proportion of the water now allocated to agriculture and to reassign it to the maintenance of river health.[3] "The new competition for water is river health versus extraction of water for economic gain, and the nation is currently engaged in substantial debate on this issue within the Murray-Darling basin" (Haisman 2003:32). In 2005, the national government indicated its willingness to step into the water market to purchase supplies that would otherwise be diverted from the river, especially if conservation measures alone do not appear sufficient to restore river flows to needed levels.

---

[3] Assuming the resumption of property rights from agriculture was accompanied by compensation payments, the total cost could be in excess of A$2 billion.

In addition, the Murray-Darling basin contains wetlands, lakes, and forests of great natural and cultural – even international – significance (Goss 2003). The Macquarie Marshes are covered by the Ramsar Agreement on Wetlands of International Importance, to which Australia is a signatory. The first step of the Living Murray initiative is focused on the restoration of adequate water supplies for six "significant ecological assets" located along the Murray, identified by the Murray-Darling Basin Ministerial Council: the Barmah-Millewa forest, the Gunbower and Koondrook-Perricoots forests, the Chowilla floodplain, the Hattah lakes, the River Murray channel, and the Murray mouth, including the Coorong and Lower lakes.

Thus, the current array of basin management issues in the Murray-Darling includes water supply allocation, limiting water use, arresting and reversing water quality degradation, and restoring and protecting ecological values.

4.2
Decentralization Process

4.2.1
Pre-reform Arrangements for Water Resource Management

Basin management has evolved and been organized in the Murray-Darling in ways that distinctly reflect the nature of Australia's federal constitutional arrangements. The Australian Constitution devolves nearly all domestic policy matters to the states, with very limited authority for the Commonwealth, or national government. Management organizations and functions have therefore never been unified at the river basin scale or uniform across states.

In addition, water resource management arrangements were, prior to reform, very centralized within each state. In New South Wales and Victoria, for example, all management and planning was carried out from Sydney and Melbourne, though water user advisory committees operated at subbasin level. There was also little attempt at cost recovery for water use or for management of headworks, irrigation schemes, or floodplain protection. Some administrative costs were recovered through, for example, license fees. Pricing for urban water supply was based largely on property tax, and cross-subsidies were common.

At basin level, the foundations for water resource management were laid down by the River Murray Waters Agreement of 1914, which set out a plan for infrastructural regulation of the Murray to ensure that riparian states, through the activities of the River Murray Commission, received their agreed allocations of water. The basic agreement was innovative for its time and lasted through a number of amendments, though it was clear by 1985 that more radical reforms were needed (see Sect. 4.2.3).

4.2.2
Impetus for Reform

Prompted by substantial fiscal problems in the late 1970s and early 1980s, the states undertook a thorough examination and reorganization of water provision and water management operations, with the aim that publicly provided services for which fees could be collected should be either corporatized (turned into governmental bodies that were financially self-sufficient – "ring-fenced" – by performing services for fees and maintaining their own assets) or fully privatized. States and territories were encouraged to undertake such reforms by the Commonwealth government, which offered financial incentives (tranche payments) for the adoption of measures consistent with an initiative known as the national competition policy, which was intended to improve public sector efficiency in Australia. Table 4.1 summarizes the management and related financing changes that took place in New South Wales and Victoria in response to these reforms.

These changes were important not only because they assisted the transition to water pricing and cost recovery practices more nearly consistent with contemporary principles, but also because they facilitated a round of other changes to state ministries. Once the construction, operation, and maintenance of infrastructure and the provision of services such as water supply had been removed from state-level departments of water resources (leaving them largely with planning and regulatory functions), a next step in most states and at the Commonwealth level was the combining of water resource departments or ministries with other natural resource or environmental departments that encompassed portfolios such as agriculture, land use planning, forestry, and fisheries. These changes have facilitated a policy shift toward integrated resource management with greater prominence being given to water quality and environmental issues.

Coupled with these state-level developments was a gradual but significant change in federal-state relationships that made it easier for the Commonwealth government to add its weight to the reform process. Although the Commonwealth government itself lacks direct constitutional authority to make and enforce water resource policy, national-level policy leadership in Australia (with respect to water and several other issues) has grown substantially since World War II, when the states surrendered their income-taxing powers to the Commonwealth, which ever since has collected income taxes on a nationwide basis and distributed the revenue back to the states (and territories), enabling it to offer financial incentives to states to conform with policy directions approved by the Commonwealth. The clearest indication of this factor in the water sector was the adoption in 1994 of the national water policy reform initiative by the Council of Australian Governments (COAG), the peak intergovernmental forum in Australia.

4.2.3
Reform Process

In 1985 discussions commenced involving the governments of New South Wales, Victoria, South Australia, and the Commonwealth to negotiate a successor to the River Murray Waters Agreement in the light of growing resource and environmental problems in the Murray-Darling basin. This process culminated in 1992 with the signing of the Murray-Darling Basin Agreement, whose charter was "to promote and coordinate effective planning and management for the equitable, efficient and sustainable use of the water, land and other environmental resources of the Murray-Darling basin" (Goss 2003). Queensland and the Australian Capital Territory were integrated into the new arrangements. The Murray-Darling Basin Commission was established to take over the transboundary water management role and was also

Table 4.1. Management functions in New South Wales and Victoria, pre- and post-1980

Function	Pre-1980	Post-1980
Water resource management	Carried out from state capital; no regional staff. All valleys had appointed water user advisory committees. No costs recovered other than administration, e.g. license fees.	Only policy and some technical specialists in state capital; regional offices in operation. Stakeholder consultation systems include catchment management organizations. New South Wales: Water pricing subject to Independent Pricing and Regulatory Tribunal. Water resource management costs recognized above operation and maintenance and renewals costs on impactor-pays basis. Victoria: Regarded as public good and funded by government.
Headworks management	All site staff controlled from state capital. No costs recovered.	New South Wales: State Water formed as internal ring-fenced business to manage headworks. Site staff and operations under regional directors, but centralized policy and standards. Victoria: Fully transferred to autonomous rural water authorities. Service committees with stakeholder input operative in both states. Full cost recovery targeted.
Irrigation scheme management	All operations and maintenance staff working in accordance with standards sourced from state capital. All schemes had appointed water user advisory committees. Limited recovery of operation and maintenance costs.	New South Wales: Fully privatized; only state management is through bulk water licenses and discharge licenses. Victoria: Fully transferred to autonomous rural water authorities. Each scheme has a user-based water services committee that oversees operation and maintenance, budgeting, and pricing. In both states operations are now self-sufficient.
Urban water supply	Function of local government (New South Wales) or specifically created local authorities, with limited consultation. Pricing based on property taxes; cross-subsidies common.	New South Wales: Water businesses of local government operating in ring-fenced fashion. Victoria: Local authorities disbanded in favor of smaller number of urban water supply state-owned corporations. Both states: Consultation now common. Pricing based on metered supplies; cross-subsidies removed.
Floodplain management	Function of local government, with state grants (New South Wales) or specially created local authorities (Victoria), with limited consultation. Both states: Owners of levees pay a license fee. Costs of local government programs covered in general property taxes plus state grants.	New South Wales: No substantial change in management. Victoria: Now transferred to catchment management authorities. Both states: Consultation now common. Levee fees still apply.

given responsibility for coordinating integrated catchment management across the whole basin. The Murray-Darling Basin Ministerial Council became the political or policymaking body (operating by consensus, with participating governments having an effective veto), and the Murray-Darling Community Advisory Committee became the community stakeholder consultative body.

In addition to operating as the executive body for implementing the Ministerial Council's decisions on basin policy and management, the Basin Commission advises the Ministerial Council on basin conditions and concerns. The commission consists of representatives from each basin government and is supported by staff; operations are funded under the cooperative agreement among the participating governments. The Community Advisory Committee advises the Ministerial Council, representing the interests and concerns of local communities and stakeholder groups throughout the basin.

River Murray Water is a ring-fenced business operation of the Murray-Darling Basin Commission; it controls the flows of the transboundary Murray River as a bulk supplier, operating infrastructure facilities on the main stem of the river to assure the states of New South Wales, Victoria, and South Australia of their flows under the provisions of the 1914 River Murray Waters Agreement (as amended).

As described in the previous section, a significant element of the reform process has therefore been the emergence of national-level leadership on water policy, with state consent. The 1994 national water policy reform initiative was revised in 2003–2004 as the national water initiative, with the following key components:

- Separation of resource management and regulatory roles of government from water service provision.
- Greater local-level responsibility for water resource management.
- Greater public education about water use and consultation in implementing water reforms.
- Research into water use efficiency technologies and related areas.
- Consumption-based volumetric pricing and full cost recovery for water services, with removal or transparency of cross-subsidies.
- New investments in irrigation schemes or extensions of existing ones are to be undertaken only after being appraised for economic viability and ecological sustainability.
- State and territory governments are to implement comprehensive systems of water allocations or entitlements, which are to be backed by the separation of water property rights from land and include clear specification of entitlements in terms of ownership, volume, reliability, transferability, and, if appropriate, quality.
- Formal determination of water allocations or entitlements includes allocations for the environment as a legitimate user of water.
- Promotion of water trading (including across state and territory borders) of water allocations and entitlements to the extent feasible within the social or physical and ecological constraints of catchments.

State and territorial achievements in the enactment of these reforms could be rewarded by the Commonwealth with tranche payments.

During 2005, a National Water Commission was established and began to meet. Its stated purposes are to promote long-range water resource planning and the coordination of state and Commonwealth policies. The commission represents another example of the recent increase in national-level attention to water resource policy issues.

4.2.4
Current Situation

Water management arrangements in the Murray-Darling basin have thus evolved from a focus on managing rivers for water quantity and security of supply, to integrated catchment management designed to maintain both water quantity and water quality and better balance water use for human consumption with that required to maintain healthy riverine systems. Governments in the Murray-Darling basin have made a conscious effort over time to adapt their arrangements to address weaknesses in dealing with emerging issues such as water scarcity and salinity, or to address instances where the arrangements have contributed to such resource degradation.

This transition is manifested in particular policy statements developed and adopted by the Murray-Darling Basin Ministerial Council, whose Natural Resource Management Strategy, adopted in 1990, explicitly embraced integrated catchment management as the basis for water resource management in the basin. The Ministerial Council and the Community Advisory Committee jointly adopted a further policy statement in 2001, Integrated Catchment Management in the Murray-Darling Basin 2001–2010: Delivering a Sustainable Future, with which all Ministerial Council strategies and actions are to be harmonized.

The states in turn have all, over the past 15 to 20 years, established forms of decentralized catchment management bodies with a mandate to advise on all aspects of natural resource management. States have, however, balked at giving these bodies too much authority, in particular the power to raise their own funds through land taxes. The arguments against more autonomous catchment management authorities that are empowered to make management decisions and raise land taxes may be grouped into three broad categories: protests from landowners, especially farmers, about the cost impost; concerns about creating a fourth tier of government; and opposition from local governments that fear encroachments upon their planning functions.

The states' reluctance to empower the catchment management bodies with autonomous revenue authority has to some degree been overcome by the national government's decision to disburse funds for natural resource management directly to properly constituted catchment management authorities under the National Action Plan for Salinity and Water Quality, and the Natural Heritage Trust program.

The institutional arrangements for governing and managing the Murray-Darling basin have been modified in substantial ways. The state governments and the Murray-Darling basin organizations have been supplemented at the subbasin level with catchment management bodies that are still developing their own roles in land, water, and natural resource management. At the national level, COAG and the Commonwealth government have become closely involved in the development of national water policy reforms and initiatives that in some respects lead and in other respects follow the integrated water resource management direction taken during the 1990s by the Murray-Darling Basin Ministerial Council. COAG's latest set of proposals, published in June 2004, would further advance a number of reforms to water licensing, water trading, and the enhancement of environmental water flows.

Table 4.2 shows the institutions that play a role in water resource management in the Murray-Darling basin, and summarizes their functions.

4.3
Application of Analytical Framework

It is now possible to return to the analytical framework (Chap. 1) and review the factors identified there as potentially related to successful development of basin-scale decentralized institutional arrangements for integrated water resource management.

4.3.1
Contextual Factors and Initial Conditions

Contextual factors and initial conditions in the Murray-Darling basin were in most respects quite favorable to the development of new institutional arrangements leading towards integrated water resource management. The level of economic development in the basin and in Australia as a whole has made it possible for stakeholders and governments to invest time and money in knowledge generation, travel, meetings, and other tasks associated with the planning, negotiation, adoption, and implementation of institutions for river basin management. There are few class, religious, or other sociocultural distinctions keeping Australians throughout the basin from being able to establish communication, share information, or make and keep agreements (where issues exist efforts are being made to resolve these – for example development of indigenous action plans in relation to natural resource management). Overall, the Murray-Darling basin was quite favorable social and economic terrain for the development of basin management institutions: its semiarid climate makes water issues significant enough to stimulate action, and the relative wealth and homogeneity of its population present few barriers to such action.

The initial distribution of resources among basin stakeholders has clearly favored irrigators in the basin, who account for more than 90 percent of water diversions. This has slowed the pace of reforms such as licensing restrictions and cost recovery pricing, with the latter driven more by national economic policy reforms than by internal basin-scale reform efforts. Indeed, national reform efforts articulated through COAG have provided helpful leverage to policy actors within the Murray-Darling basin trying to enact and implement restrictions on water diversions and the reduction of agricultural water subsidies. Current reform efforts oriented toward implementation of the Living Murray initiative have entailed several concessions toward irrigation interests. Thus, the irrigators' position has affected the shape and speed of institutional reform in the Murray-Darling basin.

4.3.2
Characteristics of Decentralization Process

If anything, the construction of basin management institutions and policies in the Murray-Darling basin has been as much a matter of integration as of decentraliza-

Table 4.2. Water resource management institutions and roles in the Murray-Darling basin

Management level	Institution	Current water management attributions
National	Commonwealth (national government)	Most water management attributions ceded to states , but the Commonwealth provides funding for natural resource management functions that support the national interest. The Commonwealth is a party to the Murray-Darling Basin Agreement, participates on the Murray-Darling Basin Ministerial Council and the Murray-Darling Basin Commission, and provides a significant portion of the funding for the commission and its activities in the basin.
	Council of Australian Governments (COAG)	Partnership of Commonwealth and state governments. Policy development and coordination, for example through the national water policy reform agenda (1994) and the national water initiative (2004).
State and substate	State governments	Full sovereign powers over land, water, and natural resources.
		State law typically vests authority for the control and use of water in a ministry, which is responsible for water rights systems, etc.
		Built and still owns and operates major dams on rivers. Initially built and operated irrigation schemes, although, with the exception of Victoria, these are now all privatized.
		Oversees, and to some degree finances, water supply and sanitation functions of local governments.
	Local governments (urban and rural)	Established and authorized by state legislation. Provide and operate water supply and sanitation infrastructure, with state government financial assistance in many circumstances. Responsible for flood protection, with state financial assistance.
Basin	Murray-Darling Basin Ministerial Council	Policymaking body for basin under Murray-Darling Basin Agreement of 1992. Composed of ministers from state and Commonwealth governments and a representative from the Australian Capital Territory; uses consensus approach.
	Murray-Darling Basin Commission	Executive body for implementing Ministerial Council decisions on basin policy and management; composed of representatives from each basin government. Also advises Ministerial Council on basin conditions and concerns. Operations funded under the cooperative agreement among the participating governments.
	Murray-Darling Community Advisory Committee	Advises the Ministerial Council, representing the interests and concerns of local communities and stakeholder groups throughout the basin.
	River Murray Water	Ring-fenced business operation of the Murray-Darling Basin Commission; controls the flows of the transboundary Murray River as a bulk supplier, operating infrastructure facilities on the main stem to assure agreed flows to states.

tion. Concerns that might arise in other countries about the ability or willingness of central government to genuinely devolve decisionmaking authority are of little consequence in Australia, where primary decisionmaking authority predominantly and initially rested at the subbasin level with the state governments. Over time, and with the cooperation and consent of the national government, the states have constructed intergovernmental arrangements to control and operate Murray River flows and then to address other issues.

Table 4.2. *Continued*

Management level	Institution	Current water management attributions
Subbasin	Catchment management boards, committees, authorities	Titles vary slightly from state to state. Mostly coordinating and advisory bodies with responsibility for protecting water quality and riparian and floodplain conditions through efforts to improve land stewardship and through actions such as riverbank protection projects and tree planting.
	Water management committees	Community-based advisory committees composed primarily of water users. Advise on water allocations, environmental flows, and in some cases flood protection, river facility operations, and water pricing.
	Irrigation companies	Fully privatized bodies to which previously governmental irrigation infrastructure assets and operations have been transferred; operate in all states but Victoria (where rural water authorities perform similar functions). They are subject to corporation law, and stand alone in all respects financially.
	Water user groups	Vary considerably in history and form. Receive some assistance from state governments because they play an important role in community participation, and in some instances even perform duties normally expected of government, e.g. creating a roster of water users and uses on an otherwise unregulated river.

Central-level recognition of basin governance and management has been complete and consistent. The Commonwealth government not only recognizes but participates in and helps to fund the basin-scale organizations such as the Ministerial Council and the Basin Commission. Through financial incentives offered to the states and to substate catchment management authorities, and through establishment of and participation in bodies such as COAG, the national government has actively encouraged the development of integrated water resource management in the Murray-Darling basin. These commitments from the national government have remained consistent across elections and changes in party control.

4.3.3
Central-Local Relationships and Capacities

Two factors in this component of the analytical framework are less than favorable to successful integrated water management at the basin level. One is the past and current system of water rights. Entitlements to the use of water are issued by states rather than by any basin-scale or larger entity, leading to some differences in the rules governing water entitlements (for example, with respect to duration, security, and transferability). Rights generally fall into three categories: licenses issued to organizations such as irrigation companies, trusts, or districts; licenses issued to individuals; and rights of riparian landowners. Overallocation of water licenses now represents one of the principal challenges of basin management. Nor has groundwater been fully integrated into the licensing system. Overall, the systems of water entitlements in the basin continue to require further reform if measures such as water trading and the protection of environmental flows are to be implemented fully.

Also, the organizations in the basin most directly associated with integrated resource management (for example the subbasin catchment management authorities) have virtually no financial resources of their own and are dependent on funding contributed to them by the state and Commonwealth governments. Although it is unlikely that this source of funding will dry up, it represents a potential area of financial insecurity.

Other factors in this category are more favorable to the successful and sustainable implementation of integrated water resource management. Especially noteworthy is the basin management participants' ability to create and modify the institutional arrangements to meet their needs and circumstances. The state and Commonwealth governments have amended and even completely replaced the agreements for the Murray River and the Murray-Darling basin during their existence, and have reconstituted the basin governance arrangements to their current structure of Ministerial Council, Basin Commission, and Community Advisory Committee. Furthermore, the participants retain the authority to make other changes in the future.

Another very favorable factor has been the extent of experience at the local and state levels with self-governance and service provision. For example, Curtis et al. (2002) have pointed out that the adoption of integrated catchment management approaches with active community participation has been helped considerably by the existence of and experience with Landcare groups since 1986, and the provision of Commonwealth funding for improved land and other resource management practices through the Natural Heritage Trust program beginning in 1988. By the time the Murray-Darling Basin Ministerial Council adopted its Natural Resource Management Strategy in 1990, there were already numerous subbasin groups addressing issues of improved land stewardship; participatory catchment management was therefore introduced into a situation already rich with social and organizational capital.

4.3.4
Internal Configuration of Basin-Level Arrangements

The existence of basin- and subbasin-level governance organizations, with firm recognition and considerable support from the state and Commonwealth governments, has created very favorable conditions for successful and sustainable development of integrated resource management. The states themselves are recognized as communities of interest within the river basin, as are a number of stakeholder communities represented on the Community Advisory Committee. Basin users and policymakers appear to have a rich array of means by which to negotiate and enter into agreements for committing and combining resources for projects and programs to improve basin conditions. Monitoring of basin conditions is performed regularly and then consolidated into a basin perspective by the commission staff, in whom considerable confidence is voiced.

Two factors about the current institutional arrangements within the basin have less certain status. The clarity of institutional boundaries has been reduced somewhat by the introduction of the relatively new catchment management bodies. Local governments within the basin are not entirely certain how the land and water management activities of these bodies will overlap or coordinate with the traditional land use regulatory authority of local governments, or with larger programs undertaken at the basin scale, such as the Living Murray initiative. These uncertainties may prove to be nothing more than temporary as each organization readjusts to the new arrangements.

Partly connected with this is the issue of conflict resolution. Arrangements exist and have been used to deal with conflicts between water users and conflicts between the states, but it is less clear how (that is, by what process or through what body) conflicts would be addressed and resolved that arose between substate and subbasin entities (such as a local government and a catchment management body), or between catchment management bodies, or between a catchment management body and a rural water authority.

4.4
Performance Assessment

4.4.1
Stakeholder Involvement

In the institutional arrangements for governing and managing the Murray-Darling basin, there are several key participants or groups of participants. Some are water user groups, such as irrigators and urban water suppliers. Others are participants with formal roles in the Murray-Darling basin institutional structure, such as the departments and ministries of the member states, the national government, and the Murray-Darling Basin Commission members and staff.

All levels of water management are now supported by stakeholder advisory groups of one kind or another. This is complete in the case of privatized irrigation schemes, where there is now no government involvement, but is also particularly well developed for integrated catchment management. The basin population has nearly 20 years' experience in such community and government partnerships and brings a highly informed and sophisticated capability to the task. Public consultation is now the norm, even for urban water and wastewater projects.

Irrigation is a very large stakeholder group in the Murray-Darling basin, but it is not a strictly homogenous group. Some irrigators are crop farmers (especially in New South Wales and Queensland) who have favored a fairly liberal granting of water licenses, and have been willing to accept less security and more variability in their water deliveries. On the other hand, there are irrigators with high capital investment in permanent plantings (orchards and vineyards, especially in South Australia and Victoria), who have favored more restrictive granting of water licenses in order to maximize the security of water deliveries to license holders. The irrigators have been more nearly united, however, in their opposition to the idea of real reductions in water licenses and diversions for environmental protection – for instance, as part of the Living Murray initiative.

State and Commonwealth policymakers promoting the Living Murray initiative might be able to reduce irrigator resistance by offering expanded water trading, so that irrigators with insufficient water allocations might be able to acquire water from those with more than adequate allocations. However, the scope of water trading is limited by the fact that it must occur within the state issuing the water license. Also, many water licenses are issued to schemes (irrigation districts, trusts, or companies) rather than to individuals, and in those cases trading among individuals depends upon the rules of the scheme rather than the rules of the state. In response to these limitations, COAG in June 2004 proposed the elimination of restrictions on interstate water trading, and on water trades out of irrigation schemes. If enacted in law, these proposals would mark a significant step toward the creation of a basinwide water trading marketplace.

Urban water suppliers in the Murray-Darling basin appear to have moved closer to full cost recovery, and their rates have encouraged conservation by their customers. Thus the prospect of reduced water diversions does not alarm urban water suppliers to the same degree it does irrigators, though concern remains as to whether resistance to further rate increases will affect the ability of urban suppliers to maintain adequate revenues while per capita and per household sales of water decline.

Another important set of participants in the basin is the state ministries related to water and other natural resources. Their paramount interests appear to be (a) maintaining state autonomy; (b) receiving Commonwealth tranche payments for following or implementing national competition policy initiatives; and (c) protecting their constituents in negotiations concerning basin policies. The combination of these motives leaves state ministries wanting to do enough to keep Commonwealth support flowing and to prevent encroachments on their autonomy, but not necessarily any more than that.

The reliance upon consensus as a decision rule in the Ministerial Council allows each state (and the Commonwealth) to block decisions or actions with which it disagrees. This may cause long delays in implementation of basin policy, or may simply encourage states to follow different policy directions. As regards the cap on diversions, for example, the council agreed to the policy, but five years later Queensland and the Australian Capital Territory had yet to determine their cap amounts, and New South Wales had calculated its cap differently from Victoria and South Australia.

A succession of Commonwealth governments have sustained participation in, commitment to, and funding for the Murray-Darling institutions. Here, the Commonwealth seems to be balancing two prime considerations: the Murray-Darling basin is economically and politically significant enough to warrant the national government's serious attention, but the basin's situation is distinct and should not necessarily be the basis for nationwide policies that would apply to other basins in Australia with different circumstances. Having a basin-scale set of institutional arrangements in which the national government can participate, but which isolates decisionmaking so decisions apply only to that basin, may satisfy the Commonwealth's combination of interests in this regard.

The Murray-Darling Basin Commission is another key participant in basin policymaking and implementation. One of the commission's key strengths is as a nonpartisan adviser, providing support without perceived favoritism or bias, and acting as a source of good neutral science and sound advice on basin policy questions. By performing these roles, the commission maintains the support of the states and Commonwealth. The commission staff are also interested in maintaining the commission's relative prominence in basin policymaking, and not having resource management issues devolved completely to state and substate actors. The commission is viewed as a premier basin organization worldwide, and this prestige is a substantial motivator for its staff.

4.4.2
Developing Institutions for Integrated Water Resource Management

Water resource management is still driven by policy elites and audit groups in each state, but all actual management is carried out at regional level in local offices with almost complete delegation for policy implementation, including water sharing. Management and operation of dams and irrigation schemes have been transferred to entities designed for completely localized day-to-day management, and for financial sustainability. In all states but Victoria, this has included the privatization of irrigation schemes and their assets into the hands of the irrigators. Urban water and floodplain management have always been local responsibilities, albeit with some central technical and financial assistance, and this has continued and intensified in both technical and financial aspects.

Goss (2003) presents five criteria for sustainable river management: stable institutional organization formally recognized by means of a treaty, law, or agreement; a technical secretariat and stable funding; a sound knowledge base; integration; and transparency and community involvement. He finds the arrangements in the Murray-Darling to have been greatly strengthened with respect to each criterion since the 1980s. Chief among the achievements cited by Goss is the adoption of the cap on water diversions, an agreement among the participants adopted by the Ministerial Council to arrest development of water use to 1994 levels, as a first step toward the restoration of river health.

Goss and others have, however, found some points of weakness in the Murray-Darling arrangements. These include:

- Frequent turnover of members occurs on the Ministerial Council and on the Basin Commission.
- The Murray-Darling Basin Commission lacks experts from outside government.
- The requirement of unanimity among commission and council members representing six governments, and for parliaments of all six governments to approve any changes to the Murray-Darling Basin Agreement, slows decisionmaking.
- There is a lack of representation on the Community Advisory Committee by some key industries affected by basin management, and many representatives on the committee are government appointments.
- While understanding of basin issues, policy, and projects is relatively strong among Ministerial Council, Basin Commission, and Community Advisory Committee members, broader stakeholder understanding of basinwide natural resource issues, decisionmaking processes, and policy matters is often poor, prompting the commission to allocate more resources to improved community participation (Goss 2003).
- Urban interests, rural towns, women, and aboriginal interests have been underrepresented on catchment management councils, at least in Victoria (Curtis et al. 2002).

- Demand management in the growing urban areas will need the kind of attention and emphasis in the near future that water use restrictions for irrigation have had recently.

4.4.3
Effectiveness and Sustainability

The national water policy reform agenda articulated in 1994, and couched in terms of a national competition policy, placed considerable emphasis on water management moving onto a sound financial footing. Economic elements of water reform policy required removal of cross-subsidies, consumption-based water pricing, new investments only if they were economically viable and ecologically sustainable, better specification of water entitlements, and the encouragement of water trading. These reforms were accompanied by institutional reforms that separated regulatory roles from service provision, required greater local-level responsibilities for management, and encouraged public education and consultation.

These reforms are advanced across the basin. Generally, both urban and rural (irrigation) water supply infrastructure now get no government funding for operations and maintenance and a very small and steadily decreasing amount of capital funding. The concept of a renewals annuity has been accepted as part of the pricing structure to ensure the long-run sustainability of the asset base.

The relationship between water resource management (including water allocation), water quality, and ecosystem health has increasingly been recognized within the Murray-Darling river basin, and a number of key programs have been adopted that address specific and general water resource problems in an interrelated fashion. These include:

- The Natural Resources Management Strategy was developed as an umbrella strategy to address many complex natural resource degradation issues on an integrated catchment management basis. The strategy focuses upon investigation and education to strengthen the knowledge and skill base. The integrated catchment management policy statement is a commitment by the community and governments of the basin to manage and use the resources of the basin in an ecologically sustainable manner.
- The Living Murray initiative (discussed earlier) involves a vision of a "healthy River Murray system, sustaining communities and preserving unique values" (MDMBC 2002:5). This seven-year program features principles of adaptive management based on detailed annual reviews of river health. The Sustainable Rivers Audit will serve as a regular assessment of river health and ecological condition. This audit also includes performance indicators for macroinvertebrates, fishes, water quality, hydrology, and physical habitat.
- Steadily increasing diversions from the rivers of the basin raised concerns that urrent levels of water use were unsustainable. The Murray-Darling Basin Ministerial Council commissioned a basinwide audit of water use, which revealed that median flow to the sea had been reduced by 79 percent from natural conditions by the high levels of diversions. The 2000–2001 review of cap implementation (MDMBC 2002) indicated that transparency in reporting concerning cap compliance was

resulting in pressure upon communities that are not in compliance, as well as their governments.[4] The review of the operation of the cap in 2002 concluded that the cap is an essential first step in achieving a sustainable basin ecosystem and that it has significantly reduced risk of environmental degradation.

- The problem of salinity is being approached on a number of fronts. The Murray-Darling Basin Commission Salinity and Drainage Strategy arose from concerns about salinity and waterlogging problems along the Murray River, with studies showing that the irrigation areas affected by high water tables could increase from 559 000 hectares in 1985 to 869 000 hectares in 2015 (MDBMC 1987). Under the 15-year strategy, states agreed to be responsible for actions taken after 1 January 1988 significantly affecting river salinity. End-of-valley salinity targets were adopted for each tributary catchment to be achieved by 2015, a market in salt credits was set up, and joint salinity interception schemes were agreed upon to manage existing saline inflows. This strategy has largely been successful in stemming salinity caused by irrigation (MDBC 1999). However, dryland salinity has continued to threaten river salinity, and the National Dryland Salinity Research, Development and Extension program has been established to further understanding of the nature of dryland salinity and possible remediation strategies. In addition, the National Action Plan for Salinity and Water Quality is a partnership plan between communities (mostly as represented by their catchment management bodies) and government to fund and manage specific actions, such as tree planting, aimed at salinity control.

- The development of a massive 1 000-kilometer algal bloom in 1991–1992 in the Darling River prompted increased concern regarding eutrophication in the basin. In 1994, the Murray-Darling Basin Commission adopted the Algal Management Strategy to reduce the frequency and intensity of algal blooms and other water quality problems associated with nutrient pollution in the basin, through a framework of coordinated planning and management actions.

In summary, the Murray-Darling Basin Commission and its stakeholders have shown commitment to understanding and taking proactive efforts to address critical issues in the basin through ongoing and reliable measurement, development of scientific knowledge, and an integrated and participatory approach, all expressed through basin- and subbasin-scale initiatives.

4.5
Summary and Conclusions

4.5.1
Review of Basin Management Arrangements

Management arrangements in the Murray-Darling basin do not represent a simple template. They are complex, they have a history that has shaped their current structure

---

[4] A region and its government are considered to be in breach of addressing water use overallocation or to not have commitment to cap compliance when breaching the cap several years running.

and direction, and they are tailored to the particular circumstances of Australian federalism and the climate and topography and basin management issues there. While certain design elements might be transportable to other circumstances (a community advisory committee, a funding formula, and so on), the overall structure has been crafted and modified over time to fit and adapt to this basin.

Its fit, its complexity, and its adaptability are among its principal strengths, and help to explain the robustness of the basin management institutions in the Murray-Darling. As noted in the preceding section, there are weaknesses and criticisms to be made of the Murray-Darling arrangements, but its successes in gaining intergovernmental cooperation and commitment, instituting mechanisms for stakeholder participation, and generating a trusted body of data about basin problems and conditions are considerable.

4.5.2
Future Prospects

Today, the individuals and organizations in the Murray-Darling basin management structure stand on the threshold of a new era, in terms of organizational arrangements and policy direction. They are incorporating subbasin catchment organizations into the framework for integrated water resource management, while leaving the basin-level organizations relatively unchanged for the time being. They are also attempting to achieve an ambitious portfolio of ecological restoration objectives, in addition to but distinct from their past focus on balancing water supply and demand for human consumption. And they undertake these efforts at a time when national-level bodies are becoming more actively involved in water policy, creating a national water policy framework into which the Murray-Darling will be expected to fit. Over the next decade, these challenges will test further the robustness of the institutions for river basin governance and management in the Murray-Darling basin.

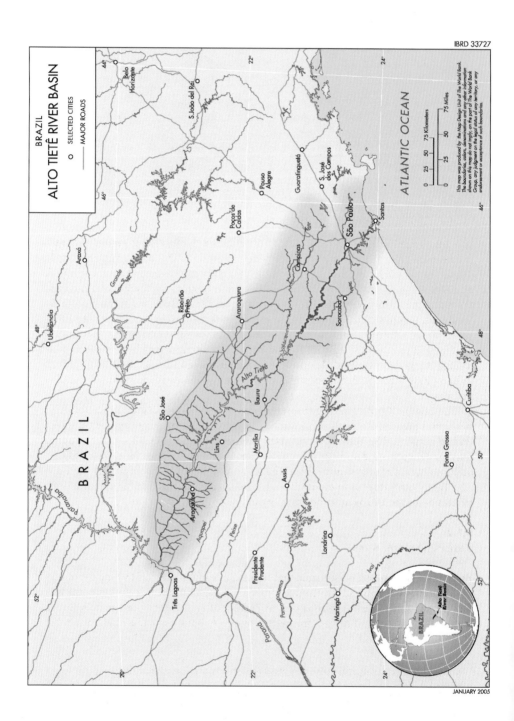

BRAZIL
ALTO TIETÊ RIVER BASIN

○ SELECTED CITIES
——— MAJOR ROADS

ATLANTIC OCEAN

0   25   50   75 Kilometers
0   25   50   75 Miles

This map was produced by the Map Design Unit of The World Bank.
The boundaries, colors, denominations and any other information
shown on this map do not imply, on the part of The World Bank
Group, any judgment on the legal status of any territory, or any
endorsement or acceptance of such boundaries.

Belo Horizonte
S. João del Rei
Pouso Alegre
Guaratinguetá
S. José dos Campos
Poços de Caldas
Araxá
Ribeirão Prêto
Araraquara
São Paulo
Santos
Campinas
Sorocaba
Ubelândia
São José
Alto Tietê
Bauru
Lins
Marília
Curitiba
Ponta Grossa
Araçatuba
Assis
Presidente Prudente
Londrina
Três Lagoas
Maringá

BRAZIL
Paraná
Grande
Aguapeí
Peixe
Paranapanema
Paranapanema
Ivaí
Paranapanema

BRAZIL
Alto Tietê River Basin

Chapter 5

Brazil: Alto Tietê Basin

R. M. Formiga Johnsson · K. E. Kemper

5.1
Background

5.1.1
Introduction

The Alto Tietê basin is located in São Paulo state, in the Southeast region of Brazil. It corresponds to the upper part of the Tietê River, which flows roughly westwards to join the Paraná River. The huge urban agglomeration of São Paulo places enormous demands on the Tietê River and other water resources, and improved management of these resources has been a subject of discussion since the 1970s, when *técnicos* (technical government officials) from São Paulo state took the lead in promoting integrated water resource management. Increased political freedom within Brazil during the 1980s encouraged the process of decentralization in decisionmaking. During the 1990s legislation on decentralized water resource management in Brazil was first approved in São Paulo state, with the Alto Tietê basin being one of the forerunners in this process.

The new institutional arrangements set up during this period have had to cope with numerous problems, including pollution of water sources and competing demands for limited supplies. The actual process itself has been beset with difficulties, including tension between state and basin-level institutions and their respective responsibilities, and continued disagreement over such issues as water use pricing. As a result the main Alto Tietê Committee has found it difficult to function effectively, and the subbasin committees have emerged as more effective organs (Formiga Johnsson 2004).

5.1.2
Basin Characteristics

Brazil is rich in water; it is estimated that about 12 percent of the world's water resources are located in the country.[1] However, distribution of these resources is extremely uneven. The Amazon river basin, which covers 48 percent of the country's territory, accounts for almost 74 percent of Brazil's freshwater resources but houses less than 7 percent of

---

[1] Surface water availability is 179 000 cubic meters per second, which reaches 267 000 cubic meters per second (18 percent of the world's water resources) if one considers the flow from neighboring countries into the Amazon, Uruguay, and Paraguay river basins. The total volume of groundwater resources is estimated at 112 billion cubic meters. All data presented in this section are from ANA 2002, IBGE 2006, and Ministério do Meio Ambiente/Secretaria de Recursos Hídricos 2006.

its population. In contrast the Northeast region, which includes most of the semiarid zone of the country, accounts for 18 percent of Brazil's territory and about 28 percent of its population but has only 5 percent of water resources. The Southeast region, with 73 percent of the country's population, 11 percent of its territory, and about 10 percent of its water resources, is the heart of Brazil's industrial, financial, and commercial economy and also has the highest agricultural production. This growth has generated increasing pressure on the region's water resources, due to conflicting demands from multiple users and the steady deterioration of water quality.

The Tietê – São Paulo state's largest river – runs 1 100 kilometers from its eastern source in the São Paulo metropolitan region to the western border of the state where it joins the Paraná River (Fig. 5.1). The Alto Tietê corresponds to the upper part of the basin, from the headwaters of the Tietê River in Salesópolis city to the Rasgão reservoir. The climate in the basin is typical of tropical high plain savannas, with a temperate summer. The average temperature in the basin is about 17.8 degrees Celsius, and ranges from annual averages of 13.8 to 24.3 degrees Celsius. Precipitation averages 1 400 millimeters per year throughout the basin.

The area covered by the Alto Tietê basin is almost coterminous with the metropolitan region of São Paulo. With a drainage area of 5 985 square kilometers (2.4 percent of the state's territory), the basin encompasses 35 of the 39 municipalities and 99.5 percent of the population of Greater São Paulo. Thirty-seven percent of the basin's territory is urbanized. Population growth in Greater São Paulo has been rapid in recent decades, and the urban sprawl continues to expand as low-income residents are continually expelled from the urban center to the city's periphery. In 2000,

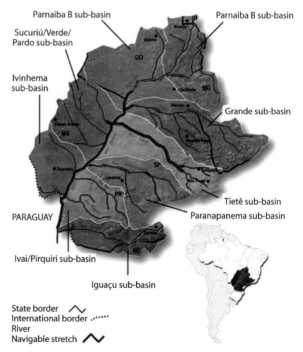

Fig. 5.1. Tietê river basin in the Paraná hydrographic region. *Source:* ANA (2002)

Parnaiba B sub-basin

Parnaiba B sub-basin

Sucuriú/Verde/ Pardo sub-basin

Ivinhema sub-basin

Grande sub-basin

Tietê sub-basin

PARAGUAY

Paranapanema sub-basin

Ivai/Pirquiri sub-basin

Iguaçu sub-basin

State border
International border
River
Navigable stretch

17.8 million people lived in the basin, and estimates are that in 2010 the population will reach 20 million (FUSP 2002a).

This massive human occupation was accompanied by the large-scale construction of water infrastructure, including dams, pumping stations, canals, tunnels, and interbasin transfers to and from neighboring basins. These projects were usually built to serve multiple purposes, especially hydropower, urban supply, and flood control. Today, the Alto Tietê basin is served by a complex hydraulic and hydrological system (Fig. 5.2). Despite this extensive water infrastructure, the water availability of the region is still very low (201 cubic meters per person per year), even lower than the semiarid regions of the Brazilian Northeast.

5.1.3
Water Resource Problems

Water resource problems in the Alto Tietê basin largely stem from the rapid urban and industrial growth during the second half of the 20th century, which has placed water resources under intense pressure and has given rise to a number of related issues.

Imbalance between Water Demand and Availability

Total water consumption in the Alto Tietê basin greatly surpasses basin water availability (FUSP 2002b; Gomes 2004), and nearly half of the current urban supply of

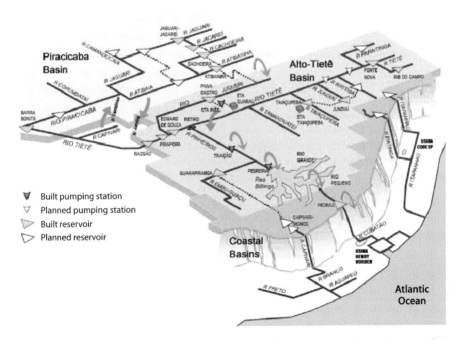

Fig. 5.2. Alto Tietê basin: Major dams and interbasin transfers. *Source:* DAEE (*www.daee.sp.gov.br*)

63 cubic meters per second is imported from the Piracicaba River basin to the north, with smaller amounts diverted from the Capivari and Guaratuba rivers (Fig. 5.2). Within a few years demand will outstrip existing water supply systems; suggested solutions include water demand management, reuse, and expanding existing system capacity, or constructing new systems linked to adjacent basins. However, conflict has arisen as the Piracicaba basin has itself undergone rapid population and economic growth, increasing significantly its own demand for water. The Alto Tietê basin plan (FUSP 2002a,b) favors a demand management approach, in order to postpone as long as possible the need to divert water from new sources.

Water Quality

According to the water quality information produced by the São Paulo State Environment Agency (Companhia de Tecnologia de Saneamento Ambiental; CETESB), the principal rivers, including the Tietê, present reasonable water quality conditions in the upstream portion of the basin. However, in the downstream portion, from the border of São Paulo municipality onwards, these rivers are classified as of extremely low quality (FUSP 2002a,b). The situation is exacerbated by inadequate solid waste collection and disposal; a large amount of garbage goes uncollected and is often disposed of in the region's rivers and lakes (FUSP 2002b). The situation for industrial solid waste is worse; in a recent study of Greater São Paulo's 39 municipalities, only 9 declared that they regulate the disposal of industrial waste (FUSP 2001). Sewage collection and treatment are also inadequate; major investments in treatment and collection network expansion were only initiated in the Alto Tietê basin in the 1990s and the situation still remains characterized by deficits in coverage (FUSP 2002b).

Water Resource Protection and Urban Expansion

Figure 5.3 shows how population growth in the metropolitan area is advancing into the three most important drinking water source systems (Cantareira, Guarapiranga-Billings, and Alto Tietê). Attempts to regulate this expansion have been unsuccessful; for example, the state Headwaters Protection Law, passed in the 1970s, was not enforced effectively, and informal residential development expanded into protected areas, particularly in the Guarapiranga basin (Kemper 1998). The law was revised in 1997 to allow certain types of land use and water management in the municipalities, permitting controlled industrialization, tourism facilities, installation of sewerage systems, and housing improvements.

Hydropower, Growing Urban Demand, and Pollution

The struggle for drinking water in Greater São Paulo has also come into conflict with the influential hydropower sector.[2] The Guarapiranga and Billings reservoirs were built

---

[2] Keck (2002) gives an overview of the problem of water supply in the São Paulo metropolitan region from a historical and political perspective.

Fig. 5.3. Urban expansion towards the major urban water supply systems of Greater São Paulo. *Source:* FUSP 2002b

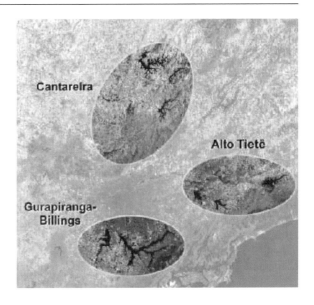

for power generation purposes in the 1920s and 1930s respectively. For decades, a large volume of water was diverted from the Tietê River and the heavily polluted Pinheiros River into the Billings reservoir for use by the Henry Borden hydropower plant (Fig. 5.2). Following increasing pressure from environmental groups pumping to Billings was suspended in 1992, and organic pollution has been significantly reduced, though plans to increase energy production by pumping water from the Tietê and Pinheiros rivers remain on the agenda of the Metropolitan Water and Energy Company (Empresa Metropolitana de Águas e Energia; EMAE).[3] The water quality of the Guarapiranga reservoir, which supplies São Paulo city, continues to worsen, even with the launching of a pollution control program (Rehabilitation of Urban Areas: Guarapiranga Project) in 1992, conducted in partnership with the World Bank.

Uncontrolled Use of Groundwater Resources

Lack of monitoring and control of groundwater use has resulted in a significant increase in groundwater abstraction, particularly by industry, with less than 2 percent of wells legalized by the Department of Water and Electric Energy (Departamento de Água e Energia Elétrica; DAEE) (FUSP 2002a,b). A 2000 study estimated groundwater extraction in the basin at 7.9 cubic meters per second, and expected it to increase to 16.5 cubic meters per second by 2010. The consequence is the lowering of

---

[3] After a major drought in 2000, which culminated in a national-level energy crisis the following year, a special license was granted to transfer an outflow of up to 4 cubic meters per second in the case of a demand for emergency power (FUSP 2002b). There are also projects for cleaning up the Pinheiros River so that it would be possible to use it as was done in the past, but now attending the environmental regulations (Agência de Bacia do Alto Tietê 2004).

the water table and a resulting increase in pumping costs, as well as the possibility of well contamination, potentially expanding to the most protected zones of the aquifer.

Urban Flooding

Urbanization in the São Paulo metropolitan region has resulted in increased imper-meability of soils and more rapid runoff, increasing susceptibility to flooding. The Alto Tietê water resource plan located 72 critical points during the floods of 1998–1999. This is, however, a municipal responsibility, outside the control of water resource policy (a state attribution). A macrodrainage plan for the Alto Tietê basin – seeking to diagnose existing and expected problems and to devise solutions from a technical, economic, and environmental perspective – was prepared in 1998 and has been updated and gradually implemented.

5.2
Decentralization Process

5.2.1
Pre-reform Arrangements for Water Resource Management

Water management practices in São Paulo state have historically been a local affair, even for the federal waters crossing it.[4] This is especially the case in the Alto Tietê basin, where all waters are under state dominion.

In technical, human, and financial terms, São Paulo's water management and environmental agencies are the best equipped in the country. DAEE was created in the 1950s and was the first to issue water use permits in Brazil. Nevertheless, it was unable to prevent or reduce the proliferation of policies affecting water use by other sectoral agencies, with little coordination. In addition CETESB, created in the 1970s, is widely respected for its technical competence; however, its command-and-control regulations have been mostly limited to the state's largest industries and worst polluters, and some other players, for example water and sanitation companies, have been subject to much weaker regulation and enforcement.

Thus, despite their resources, the São Paulo state management agencies were still unable to control the speed with which water resources in the Alto Tietê basin were being appropriated and used. In addition, there has been little coordination and integration between DAEE and CETESB to manage water quantity and quality, and between those agencies and many others directly involved with problems of water use and planning in Greater São Paulo, such as the Metropolitan Planning Agency for Greater São Paulo (Empresa Metropolitana de Planejamento da Grande São Paulo; EMPLASA).

---

[4] The distinction between federal and state waters is defined in the 1988 Constitution. The federal government has jurisdiction over waters that cross state or international boundaries. The water located entirely within the territory of a single state, as well as groundwater resources, are in state domain, except when they are used by federal infrastructure projects.

The system in the Alto Tietê basin prior to reform could therefore largely be characterized as (a) compartmentalized (quantity separate from quality, surface water separate from groundwater resources); (b) centralized finances, planning, and decisionmaking at the state level, since the municipalities, private users, and civil society had no say in how to manage the water resources; and (c) inadequate, considering the insufficient technical, administrative, and financial resources available for planning, control, and enforcement activities when compared to the complexity and magnitude of the problems (Formiga Johnsson 1998).

5.2.2
Impetus for Reform

The impetus for reform of water resource management in the Alto Tietê basin emanated from a number of factors operating at national, state, and basin levels. At the national level, the technical water resource community – led in large part by técnicos from São Paulo – began in the 1970s to promote integrated water resource management within river basins. These ideas received encouragement during the 1980s, when political democratization signaled readier acceptance of decentralization and civic participation in policymaking.

Brazil was thus well placed to be a forerunner in adopting a new water resource management system in consonance with the principles recommended by major international charters and organizations such as the Dublin Statement of 1992 and the World Bank in 1993. These principles include the integration of sectoral policies, the decentralization of management to river basin level, the participation of stakeholders, and the concept of water as an economic good.

São Paulo was considered a state in which reform of water resource management was both necessary and feasible. The state in general, and the Alto Tietê basin in particular, was characterized by high levels of financial and industrial development and output, high average per capita income, and relatively sophisticated institutional development. On the other hand, rapid economic and urban development threatened the water supply within the region, and current arrangements were inadequate to deal with this threat.

However, positive processes and factors were at work increasingly during the 1980s. The water management institutions to a large extent favored reform and had the capacity to implement it. Social movements within the state and the Alto Tietê basin were becoming increasingly vocal in demanding new water policies and updated institutions capable of dealing with the water management challenge, as it became increasingly clear that integrated water resource management was essential for the sustainability and rational use of water resources in the basin.

5.2.3
Reform Process

Stimulated by the factors described in the previous section, São Paulo state became the first in the country to define and adopt a new decentralized state system for water resource management, formalized as Law 7.663 of 1991 (Table 5.1). The intention was to create regulatory bodies at state and river basin levels that would then

Table 5.1. Main elements of São Paulo Water Law (1991) and complementary legislation

---

Objective of the state water resource policy

- Reliability of water availability for current and future generations, at the desired level of water quality

Principles

- Integrated water management, with the river basin as the planning unit
- Water as a finite and fragile resource
- Water as an economic good
- Decentralized and participatory management

Organizations

- State Water Resources Council (Conselho Estadual de Recursos Hídricos; CRH): Deliberative body, with stakeholder participation, in charge of supervising and regulating the state water resource management system
- Coordinating Committee for the State Plan for Water Resources (Comitê Coordenador do Plano Estadual de Recursos Hídricos; CORHI): CRH's executive secretariat; technical body in charge of elaborating the state water resource plan and promoting institutional integration among the state water-related institutions
- State water management agencies: In charge of water use and pollution control, and implementing the water resource management system
- River basin committees: Deliberative stakeholder bodies with decisionmaking and regulatory powers
- River basin agencies: Technical and administrative arms of river basin committees

Instruments

- Basin water resource plans, state water resource plan[a]
- Bulk water use permits
- Bulk water charges[a]
- Classification of water bodies according to predominant use and water quality standards
- State water resource information system[a]

---

[a] These instruments were first introduced by the state law. The remaining tools already existed, but were either not functioning as planned or their structure and functioning was modified significantly by the law.

define the details of how the management instruments foreseen in the legislation (for example water permits and charges) would operate. In the Alto Tietê basin this process was marked by two distinct processes: (*a*) decentralization from the state to the basin level, which occurred with the creation of the Alto Tietê Committee in 1994 and, more recently, its water agency; and (*b*) further decentralization pushed by the Alto Tietê Committee within the basin in 1997–1998, which resulted in five subcommittees at lower territorial levels.

Making the river basin the basic unit in the organizational structure was largely based on the French system. The center of gravity of the new system would be the river basin committee, whose formal functions would include setting guidelines and approving river basin plans; proposing pricing criteria and values for water tariffs and a program for allocating water tariff proceeds; and integrating the decisionmaking and programs of water-related institutions working in the basin. Basin agencies would be the executive arm of the committees, providing technical support and implementing their decisions. Bulk water charges would give financial autonomy to the new basin institutions. A fund to finance water management was also established – the State Fund for Water Resources (Fundo Estadual de Recursos Hídricos; FEHIDRO). The 1991 Water Law had called for the immediate creation of the Alto Tietê Committee, but it was only formally established in November 1994.

Committee governance was divided into four bodies: the executive board, the assembly, the executive secretariat, and the technical chambers. The 48 seats of the committee assembly are divided equally among representatives of three sectors: municipal government, state government agencies (including public water users), and organized civil society groups (including those representing private water users). This composition is quite different from the later 1997 federal law, in which users are grouped together and must occupy exactly 40 percent of the seats and civil society at least 20 percent. For this reason, many have called the São Paulo model overly state based.

To help deal with the complexity of such an intensely urbanized and industrialized region, in 1997 the committee proposed a division of the Alto Tietê basin into five hydrologically based subbasins under the aegis of five subcommittees (Rocha 2002).[5] This division was designed in 1993–1994 but was only implemented in 1997 in the context of the revision of the Headwaters Protection Law, which called for the elaboration of specific legislation for each subbasin of the Alto Tietê. For this reason, although they also have the same attributions of the central committee, the main responsibilities of the subcommittees are the regulation and implementation of headwater conservation, protection, and recuperation policy at the local and regional levels. Their deliberations must be submitted for approval to the Alto Tietê Committee assembly, which is responsible for promoting the integration of subbasin policies.

The Alto Tietê Committee created its basin agency in 2001. However, this agency is poor in technical, financial, and institutional capacities, while DAEE remains the committee's executive arm in charge of technical and administrative support. Fully working basin institutions have yet to be created in the Alto Tietê basin, mainly because the financial vitality of these bodies remains very limited. Between 1994 and 2003, FEHIDRO allocated R$21 million (US$9.8 million)[6] to the Alto Tietê Committee. This is only enough to minimally sustain these basin bodies until the management system is fully operational.

Water charges were supposed to be introduced in 1995–1996, but lack of political will to approve complementary legislation blocked the process for a long time. The state Water Charges Law (Law 12.183/2005) was finally adopted in March 2006. Now every committee can start its own process of adopting the water charge system in accordance with the state legislation. The committee's proposals are also under the regulation of the State Water Resources Council (CRH).

Other management instruments have advanced further. DAEE has actively issued water permits in recent years, especially for withdrawal and consumption. The

---

[5] Cotia-Guarapiranga (1997), Juqueri-Cantareira (1997), Billings-Tamanduateí (1997), Pinheiros-Pirapora (1998), and Tietê-Cabeceiras (1998).
[6] 1 US dollar = 2.15 Brazilian reals (March 2006).

number of water permits issued in São Paulo state as a whole ranged from 215 in 1994 to over 4000 every year since 2002. Most of the 23 state committees also have their own basin plan, and several updates have been made of the state water resource plan since 1991.

All in all, the São Paulo management system can be characterized as reasonably advanced, even though the rhythm of implementation has been much slower than the initial process of approving the Water Law and creating the basin committees.

5.2.4
Current Situation

Table 5.2 indicates the main institutions for water resource management in the Alto Tietê basin, and their principal functions. In practice the National Water Agency (Agência Nacional de Águas; ANA) and other federal institutions exercise little power in the basin compared to the state-level Energy, Water Resources, and Sanitation Secretariat (Secretaria de Energia, Recursos Hídricos e Saneamento; SERH). This body coordinates the state water management system, including DAEE and the Basic Sanitation Company of São Paulo State (Companhia de Saneamento Básico do Estado de São Paulo; SABESP). In a similar way the State Secretariat for the Environment (Secretaria de Estado do Meio Ambiente; SMA) coordinates the activities of CETESB. Other state-level government organizations include CRH, CORHI, and FEHIDRO.

The Alto Tietê Committee was set up to integrate water resource management in the basin, and has a high proportion of state government representation. It is funded principally by FEHIDRO, since the Alto Tietê Basin Agency remains unable to provide technical and administrative support to the committee, with DAEE performing these functions. The subcommittees of the Alto Tietê basin are currently proving to be more effective organizations, although they still do not yet have their basin agencies.

5.3
Application of Analytical Framework

5.3.1
Contextual Factors and Initial Conditions

In the Alto Tietê basin and in São Paulo state more generally, the context seemed favorable to the development of a decentralized and integrated water resource management system. The state is by far the most important industrial and financial center of Brazil, generating almost 32 percent of the country's gross national product in 2003. Its average per capita income (2003) of R$12 619 (US$5 869) is well above the national average of R$8 694 (US$4 044) (IBGE 2006). It was also reasonable to expect that the strongest advances in promoting decentralized stakeholder models of water resource management in Brazil could take place in the state that had the richest, best equipped, and most experienced water management institutions.

In social terms, both the Alto Tietê basin and the state as a whole was fertile terrain for reform: motivated by worsening water-related problems, a result of intense urbanization and industrialization, social movements demanding improved water policies emerged in the mid-1980s, especially in the Alto Tietê and Piracicaba basins.[7] DAEE técnicos saw water reform based on a sound water resource policy as necessary if DAEE were to carry out its formal water management attributions.

The initial distribution of resources among basin stakeholders also seemed to favor reform. The main users are the urban water supply companies, which face serious problems of supply in the face of growing demand. On the other hand, the irrigators – traditionally the water users that are most resistant to change, and especially to water charging – use an almost insignificant amount of water in the Alto Tietê basin, although they are a major user at the state level.[8] The industries are the second largest user in the basin.

Despite this favorable context, the political will to advance the change has proved insufficient to overcome the resistance and fears of the stakeholders discussed above. The political and environmental complexities of the Alto Tietê basin seem to make it particularly difficult to implement practices involving integration and participation in decisionmaking.

5.3.2
Characteristics of Decentralization Process

As previously noted, the decentralization process in the Alto Tietê basin occurred in two stages: decentralization from the state to the basin level, and further decentralization within the basin, resulting in the setting up of five subcommittees at lower territorial levels. While the devolution of authority and responsibility from the state level to the basin level was desired by both state government and local stakeholders, there is no agreement about the extent of this decentralization, particularly regarding the financial autonomy and capacities of the river basin bodies.

The need to decentralize within the basin, however, had consensus from the earliest stages of mobilization for the creation of the Alto Tietê Committee (Rocha 2002). Although the Alto Tietê is a small basin in physical terms, state and local stakeholders recognized that the complexity of such an intensely urbanized and industrialized region required smaller scales for management. Some have disagreed with the methodology used to define the five subcommittees that were created, because they were not based on purely hydrological criteria. But few question either the need to create complementary deliberative bodies at lower levels or the fact that basin management participants should be allowed to create and modify institutional arrangements according to their needs and circumstances. The fact that there is no conflict between

---

[7] An in-depth analysis of the societal movements, and of the São Paulo system's construction and its interface with the national water management system, can be found in Formiga Johnsson 1998.
[8] While irrigation in the basin uses only 2.56 cubic meters per second, the sector uses 101.56 cubic meters per second in São Paulo state as a whole.

Table 5.2. Main institutions for water resource management in the Alto Tietê basin

Management level	Institution	Current water management attributions *Financial aspects*
Federal government *Water legislation, hydropower, management and control of federal waters*	ANA National Water Agency	ANA (and other federal institutions) have little influence in the Alto Tietê basin, except regarding hydropower issues and interbasin transfers from federal rivers. *Administrative and financial autonomy, funded by the federal budget and by royalties from the hydropower sector*
State government *Management and control of state waters*	SERH Energy, Water Resources, and Sanitation Secretariat	Establishment of state water resource policy. Coordination of the state water management system, including DAEE and SABESP. *Funded through the state budget*
	DAEE Department of Water and Electric Energy	Key institution for the conception and implementation of the state water resource policy and system and is the main technical support for urban flooding control, drainage infrastructure, and other water-related projects of São Paulo municipalities. In the Alto Tietê basin, it provides technical and administrative support to the main committee and the subcommittees. *Funded through the state budget*
	SMA State Secretariat for the Environment	Among its many attributions, SMA is responsible for the establishment of state environment policy and coordination of the state environment management system, including CETESB. *Funded through the state budget*
	CETESB São Paulo State Environment Agency	Among many attributions related to environment, CETESB is responsible for issuing environmental permits and, through its various offices, monitors and enforces compliance of all kinds of polluters. *Funded through the state budget and environment* multas *(penalties)*
	SABESP Basic Sanitation Company of São Paulo State	Provides urban water supply and sanitation services to São Paulo city and most municipalities of the Alto Tietê basin, and is also the largest user in the basin. *Funded through water tariffs and the state budget*
	CRH State Water Resources Council	Deliberative body, with stakeholder participation (state, municipalities, civil society), in charge of supervising and regulating the state water resource management system. CRH relies on the technical expertise and administrative support of CORHI.
	CORHI Coordinating Committee for the State Plan for Water Resources	CORHI is the technical body of the system formed by DAEE (coordinator), CETESB, and the secretariats of water resources and the environment, and is the executive secretariat for CRH. It coordinates the revision of the state water resource plan every four years and promotes institutional integration among all the state water-related institutions.
	FEHIDRO State Fund for Water Resources	São Paulo's water management fund, to be used mostly for projects and activities approved by the basin committees. Since its establishment in 1994 only royalties from the hydropower sector have been entering FEHIDRO, totaling about R$208.6 million (US$97 million).

Table 5.2. Continued

Management level	Institution	Current water management attributions *Financial aspects*
River basin	Alto Tietê Committee *Formal institution under state jurisdiction and regulation*	Tripartite composition: municipalities, state agencies, and organized civil society groups. Main attributions include integrating water-related institutions and programs in the basin; approving river basin plans; proposing pricing criteria and values for water charging; approving plans related to the Headwaters Protection Law of 1997; and proposing the (re)classification of water quality for sections of rivers. *Funded by FEHIDRO, with occasional financial support from its members*
	Subcommittees of Alto Tietê basin *Formal institutions submitted to the Alto Tietê Committee's regulation*	Same tripartite composition as the main committee. In addition to the main attributions of the main committee, the subcommittees are responsible for the regulation and implementation of headwater conservation, protection, and recuperation policy at the local and regional levels. *Funded by FEHIDRO, with occasional financial support from its members*
	Alto Tietê Basin Agency	In charge of providing technical and administrative support to the Alto Tietê Committee. Among its attributions, the agency should be responsible for collecting water charges and elaborating an investment plan for its utilization, though in reality it has very limited capacities. *Funded by São Paulo municipality and by occasional funds from public and private institutions*
Municipal government *Constitutional powers over land use, urban drainage, and water supply and sanitation*	Municipal secretariats of São Paulo city and other municipalities	Shares responsibility with state government for urban drainage and local environmental issues. São Paulo city is a member of the Alto Tietê Committee and all subcommittees in the basin. *Funded by the municipal budget*
	Municipal water and sanitation services	Only a few municipalities are not supplied by SABESP and have their own local urban water supply services. *Funded by water tariffs and municipal budgets. Limited recovery of operation and maintenance costs. No financial or technical assistance from state*

the responsibilities of the central committee and the subcommittees demonstrates that the decentralization that occurred is satisfactory for both sides (the center and the local levels), even though difficulties in coordination do exist.

5.3.3
Central-Local Relationships and Capacities

Prior to reform, São Paulo state already had well-equipped management institutions with a highly qualified technical corps, so decentralization has largely entailed developing a culture of integrated management among technical state officials and building capacities in new management practices involving shared decisionmaking.[9] A decade after the process began substantial improvements can be observed, including increased integration of information systems and actions, and the elaboration of water resource plans at both the state and basin levels that reflect more comprehensive and higher-quality understanding of water problems.

Above all, considerable advances have occurred in water use controls through the implementation of a new water permit system, even though monitoring and control is still not systematic (Baltar et al. 2003). The water permit has gradually become a strategic element in water resource management and control, and will take on even greater importance with the implementation of bulk water charges. Volumetric charges will not apply to actual use but rather to the size of water use permits, in order to promote a greater association between permit requests and actual needs. In the absence of a water pricing mechanism, users often request permits for volumes above their intended usage.

Until the system is fully operational, other capacities that eventually should develop at the river basin level remain in the purview of state agencies. The main state institutions – SMA, CETESB, and particularly DAEE – have been providing technical and administrative support to the basin committees. FEHIDRO has been providing small but regular amounts of money to all basin committees. The Alto Tietê basin receives an annual average of less than R$2.3 million (US$1.07 million), derived from hydropower sector royalties, which is divided between the main committee and the subcommittees (one third and two thirds, respectively).[10]

In short, the advances in state water management capacity have been considerable and in some cases crucial for the survival of the basin committees in this transitory phase. However, tensions and problems exist between the central authorities and the local bodies and, as the Alto Tietê case exemplifies (see Sect. 5.3.4), basin committees are not always effective. Indeed, the São Paulo water resource management system as a whole has been showing signs of breakdown in the face of the state government's incapacity to make it fully operational, especially by implementing bulk water charges. However, the recent approval of the Water Charges Law in December 2005 raises hopes for substantial advancements.

5.3.4
Internal Configuration of Basin-Level Arrangements

All of the new institutions defined in the Water Law have been formally implemented in the Alto Tietê basin. However, the Alto Tietê Committee has yet to become a

---

[9] São Paulo is one of the few exceptions to a recent phenomenon in Brazilian state government, in which most qualified técnicos in water management are consultants, usually paid through international agencies. A disadvantage of this trend is that states have less incentive to build their own qualified professional corps. Only recently have these states developed some of their own institutional capacity, a result of current reforms (Garjulli 2001b).

[10] A study conducted by Alvim (2003) showed that the projects are relatively well distributed geographically among the areas of influence of the various decisionmaking bodies: 16 percent were destined to projects proposed by the main committee; 26 percent to the Cotia-Guarapiranga subbasin; 18 percent to the Tietê-Cabeceiras subbasin; 15 percent to the Billings-Tamanduateí subbasin; 13 percent to the Juqueri-Cantareira subbasin; and 12 percent to the Pinheiros-Pirapora subbasin.

forum for effective decisionmaking, and its performance has tended to vary with the effectiveness of its top officials (Keck and Jacobi 2002).[11] For example, during one dynamic period following 1997, the committee was able to promote a broad integrated urban drainage policy for Greater São Paulo. The committee also cosponsored a two-year process of public hearings and debates on the revision of the Headwaters Protection Law (Keck 2002). More typically, however, state institutions, including SABESP and DAEE, tend to make major water-related decisions without going through the committee. Gaining influence over state programs constitutes the main challenge for all basin committees in Brazil, especially those with little or no capacity for implementing a water pricing system.

After 2001 the situation apparently worsened,[12] and the main committee weakened to such an extent that it essentially only discussed the allocation of FEHIDRO funds. The basin agency, which does not have much technical capacity, took on the role of the committee for other issues, actively participating in debates about water management questions affecting the basin; and DAEE provided technical and administrative support to the main committee (which should be the role of the water agency) and will probably continue to do so until the agency becomes financially independent with the institution of water pricing.

The subcommittees are generally considered more dynamic and effective than the main committee (Alvim 2003; da Cunha 2004; personal interviews in June 2004). The most important role of the subcommittees is to deal with one of the most serious water-related problems of the basin: making water resource protection and urban expansion compatible through the implementation of the state Headwaters Protection Law of 1997. This law reflects the new approach within São Paulo that attempts to integrate water quantity and quality, linking the management of water to its environmental aspects, especially water pollution and land use. Despite this promising start, however, the implementation of such policies is likely to face significant difficulties, since reaching the proposed goals depends on the capacity and will of municipal authorities to improve their urban regulations so as to guarantee the control and monitoring of land use in the subbasins.

5.4
Performance Assessment

5.4.1
Stakeholder Involvement

Water reform in São Paulo state and in the Alto Tietê basin has considerably changed the political scenario for water resource management at both the state and basin levels. Stakeholders that in the past were entirely excluded from decisionmaking – particularly municipalities, private water users, and civil society – have come onto the political scene and important steps towards further decentralization have been taken, though tradi-

---

[11] The study conducted by Keck and Jacobi in the context of the Watermark Project describes in length the dynamics of the Alto Tietê Committee between 1997 and 2001, within the broader institutional context and the process of social mobilization and organization.
[12] Keck and Jacobi (2002) already observed a significant decline in committee dynamism in 2001.

tional power positions have been slow to change and levels of participation continue to vary. The state government, through DAEE and CETESB, tends to dominate agenda setting and discussion outcomes in the main committee and subcommittees (Alvim 2003). Da Cunha's (2004) analysis of social networks in the Guarapiranga and Billings subcommittees notes that actors from the state and municipal governments interact strongly while civil society representatives are clearly marginalized from decisions.

Since the beginning of the water reform, two main issues have generated conflict and are of crucial importance to the issue of stakeholder involvement: first, the degree of decentralization and the related issue of whether the revenues from water charges (once they are implemented) should be administered at state or basin level; and second, delays in the implementation of a pricing system for state waters have slowed the advancement of the overall water management system. The interaction between key stakeholders on these issues has been a crucial element of the decentralization process.

The first issue generated intense discussion when the final version of the Water Law was being drafted in the early 1990s. DAEE has always held that it should be responsible for collecting bulk water charges and for centralizing part of the revenues at state level, an arrangement that would have severely limited decentralization since the basin committees would essentially be consultative bodies, without their own administrative, technical, or financial capacities (Formiga Johnsson 1998). However, the increasingly active stakeholders mentioned above (municipalities, private water users, civil society) succeeded in introducing the river basin agency into the Water Law, making it easier to guarantee the return of all revenues to the basin of origin[13] and giving a basin-level institution some degree of financial and technical autonomy. The specific state legislation on basin agencies (Law 10.020/1998) defined them very much as they are in the French system (*agences de l'eau*), in that they operate as the executive arm of the basin committees, providing administrative and technical support, elaborating basin plans, charging for water use, and designing investment plans for spending the revenues from water tariffs.

After that law clarified the water agency issue, the power struggle around centralization versus decentralization moved to the discussion of the 1998 draft law on bulk water charges, which was approved only in December 2005 and which, finally, defined the remaining issues around the water charging system, including that all proceeds will return to the basin, a decision of crucial and basic importance to the effectiveness and autonomy of basin-level institutions.[14]

Several public stakeholders remained interested in the status quo. The main actor pressuring for water reform, DAEE, had fought against a water pricing bill that called for the full decentralization of the allocation of proceeds, as was finally approved, and has always called for the centralization of some of the charge revenues to fund

---

[13]The São Paulo Law 7.663/1991 defined that up to 50 percent of the proceeds could be used in other basins with the previous agreement of the committee of the basin of origin.

[14]According to the law, (a) 7.5 percent of proceeds can finance the state system; and (b) up to 50 percent can be used in basins other than the basin of origin, provided the respective committee approves the transfer. This is an important exception since many committees belong to the same hydrographic basin.

strategic investments defined in the state water resource plan. It is now widely expected that DAEE will fight to be the water agency of many committees in order to retain its influence over the committees' agenda and over the use of the water charge proceeds by the committees. By law, the committee can decide to have its own basin agency or to have DAEE performing the same functions.

Although it has never taken a formal position against water charges, the state Secretariat of Finance fears the political impact of creating a new "tax". SABESP, seeing itself more as a user-payer than as a potential beneficiary of water tariff implementation, was also known to be reticent towards the tariff system and has succeeded in imposing a charge reduction of 50 percent for the municipal sector until 2010.[15] The National Association of Municipal Sanitation Services (Associação Nacional dos Serviços Municipais de Saneamento; ASSEMAE), however, had accepted the proposal of water charging at an early stage, on the condition that the funds return to the basin and that the committees have the autonomy to decide how they are allocated (Miranda-Neto and Marcon 2000).

Generally, local stakeholders, private water users, civil society, municipalities, and basin committees strongly favor the total decentralization of revenues to the basin of origin, even if water pricing alone will not resolve needs for investments in water protection and restoration. Industrial and agricultural users particularly have made it clear that they do not want to pay water charges; however, if water charges are implemented, they will only be palatable for these groups if the proceeds return to the basins from which they are generated and thus benefit those who pay.

In summary, until December 2005 the main political issues related to decentralization had to do with the final destination of water charging revenues and, in turn, the specific nature of the basin agency. The mobilization associated with reform was only sufficient to overcome the state government's inertia and lack of political will on this issue a decade and a half after the passage of the state law (1991) and the creation of the first basin committees (1993–1994).

5.4.2
Developing Institutions for Integrated Water Resource Management

Considering the fact that the state government rarely discusses major projects with the committee, it is clear that devolution of state authority over water management issues to the Alto Tietê committees has been very limited thus far. The problem is not just resistance to devolution on the part of the state government: the main committee has also failed in recent years to engage in the most important issues for the basin, as noted in the previous section.

Decentralization within the basin (from the main committee to the subcommittees) has, however, been very effective in the Alto Tietê basin. The subcommittees are making important local decisions and these have been systematically confirmed by the main committee, even though improvements are still necessary in this relationship (Alvim 2003; Rocha 2004; personal interviews in June 2004). In general, the

---

[15]These conclusions are based on Formiga Johnsson 1998, Carmigmani 2000, and interviews with stakeholders in São Paulo state during field visits in June and October 2004.

subcommittees continue to be more active, dynamic, and stable, despite the ongoing difficulties afflicting the main committee. This strongly suggests that decisionmaking over water management issues in very dense and urbanized areas may be more appropriate at lower levels.

This observation is especially important in the Alto Tietê, a hydrographic region that encompasses only the very upper part of the Tietê basin. At the time it was created, the reformers believed that a region of that size would be small enough to ensure that a committee could face the magnitude of São Paulo city's problems. The complex array of problems and the wide range of vested interests has, however, proved extremely difficult to coordinate at basin level, and the state government's failure to include the committee in decisionmaking has added to the perception that it is less politically relevant.[16]

Whatever the degree of decentralization, financial arrangements that make sufficient funds available to sustain the system are essential. The funds available for the Alto Tietê basin – an annual average of less than R$2.3 million (US$1.07 million) – have been enough to provide minimal financial sustainability for all the basin's committees. The basin's water management needs are, however, estimated at about R$5.3 billion (US$2.15 billion) for 2001 to 2010. Of this total, 97.2 percent refers to services and infrastructure and 2.8 percent to management activities (FUSP 2002a).

Preliminary simulations made by São Paulo state (CORHI 1997) have, however, noted that the Alto Tietê has the greatest potential among São Paulo basins to generate revenue through charging for water use. The São Paulo bulk water charge system proposed in the state law (Law 12.183/2005) differs from the French system in that it directly links charges to the volume of water for which the user has a water use permit, but in other respects it is similar: basic charges per unit are applied for type of use (withdrawal, consumption, effluent discharges) with a multiplier coefficient to take local specificities into account, such as type of water source (surface or groundwater), seasonality (wet or dry seasons), and location (water source protection areas, for example) (Table 5.3).[17] The final charge (basic charges × coefficients) should have a ceiling (maximum charges) to avoid having a significant impact on users. All water users should be charged: first, industrial and municipal; from January 2010, irrigation and others.

Studies carried out by CORHI in 1997[18] predicted the potential annual revenue of water use and pollution charges in the Alto Tietê basin to be R$178 million

---

[16]Despite these considerations, it is still important to avoid losing sight of the hydrographic region as a whole, as has occurred in other committees that share the same river basin in São Paulo state (personal communication of Ney Maranhão, coordinator of the studies for the São Paulo state water resource plan 2004–2007, in November 2004).

[17]Note that pollution is charged according to the pollution load. This is quite different from using effluent dilution (the volume of water needed to reach the concentration of pollutants set by standards in the water body where effluent are released) as occurred in a simplified form in the Paraíba do Sul river basin, and as has been proposed by several Brazilian studies (see, for instance, Laboratório de Hidrologia e Estudos do Meio Ambiente/COPPE/Universidade Federal do Rio de Janeiro 2004).

[18]The simulation for the Alto Tietê considered that industrial users, urban users, and irrigators would pay. All the coefficients were equal to one.

Table 5.3. Maximum bulk water charges for industrial and municipal users in São Paulo state. *Source:* São Paulo state's Law 12.183/2005 and Decree 50.667/2006

Water use	Unit	Maximum charges[a] R\$ (US\$)
Withdrawal	Cubic meter	0.01 (0.0046)
Consumption	Cubic meter	0.01 (0.0046)
Effluent discharges:		
Biological oxygen demand (BOD)	kg BOD	0.03 (0.0140)
Other effluent quality parameters to be defined by the basin committees two years after implementation of the pricing system	Variable	The maximum charges of all quality parameters together are three times greater than the maximum charges for withdrawal and consumption

[a] Water and sanitation companies, except those of the Alto Tietê and Baixada Santista, will pay only 50 percent of the total amount until the end of 2009, if the same amount is invested in wastewater collection and treatment.

(US\$82.8 million) (CORHI 1997).[19] However, this estimation was based on higher assumed rates than those finally approved by the law and its complementary legislation. Moreover, it is estimated that over a third would go to the other basins that provide water to Greater São Paulo: the Baixada Santista and, above all, the Piracicaba-Capivari-Jundiaí. The law also requires that another 7.5 percent be used to finance the state system. Therefore, the potential revenue that could be used for investments in the Alto Tietê basin is estimated at under R\$100 million (US\$46.5 million) a year, based on the assumption that all potential user-polluter-payers will comply with the new legislation. This means that revenues from water charges would still be insufficient for ensuring the basin's financial self-sufficiency, covering only less than 20 percent of annual investment needs.[20]

In conclusion, even though the water charges will not guarantee full financial self-sufficiency, they will constitute a strategic foundation for the decentralized management system in the Alto Tietê basin.

5.4.3
Effectiveness and Sustainability

The issue of establishing a financing mechanism that will be sustainable in the long term has been discussed in the previous section, in relation to the institutions involved. Other criteria by which effectiveness can be measured are discussed below.

---

[19]Two other simulations were carried out later (1999 and 2001): one reached the same estimated revenues collected, but the second estimated only about R\$110 million (US\$51.2 million) per year (FUSP 2002a).

[20]Nevertheless, the needs estimate took into account some sanitation and drainage projects that have funding already guaranteed by executing agencies – DAEE, SABESP, municipal water and sanitation companies – in some cases with the participation of international agencies.

Although it is still too early to allow for a complete analysis of the long-term physical impact of decentralization in the Alto Tietê basin, a series of changes can be noted.

Decisionmaking Regarding Investments

Given the small amount of funds available, the Alto Tietê Committee's strategy has been to give priority to nonstructural projects (for example rationalization of water use, demand management, capacity building, environmental education), leaving traditional state agencies (SABESP, DAEE) to carry out large-scale infrastructure projects.[21]

Water Permits

DAEE has intensively and progressively issued water permits in São Paulo state since the beginning of the reform. A recent World Bank study (Baltar et al. 2003) emphasized that knowledge about water resources in São Paulo state is strong, thanks to DAEE's institutional capacity. To support the water permit system, DAEE has about 150 field officers and 40 others in the São Paulo headquarters, much more than any other state in Brazil. However, the study also pointed out that there is still a need for regular monitoring and control, which occurs only in the case of complaints by third parties.

Water Resource Protection, Urban Land Use, and Environmental Policy Integration

The Alto Tietê Committee and subcommittees will probably face difficulties in approving and implementing the specific laws and programs they are elaborating in conjunction with the Headwaters Protection Law of 1997. However, the importance of this initiative goes beyond the fact that a major problem has been openly discussed and politicized. This new policy is one of few cases in Brazil where water policy is integrated with more traditional environmental policies. It is also the first major attempt to integrate land use, urban development, and water resource management in heavily urbanized areas in the country. The recognition of these efforts can be found in the Water Charges Law, which calls for the approval of these specific laws and, above all, defines that at least 50 percent of the total collected revenues in the Alto Tietê basin should be invested in the conservation, protection, and restoration of its headwater areas.

---

[21] The most important of these investments is the Tietê River Pollution Control Project financed by the Inter-American Development Bank, which is considered the largest sanitation project in Brazil. Its overall objective is to improve the environmental quality of the Tietê basin in the São Paulo metropolitan region through four components: (*a*) cleanup of the Tietê River; (*b*) sewer systems; (*c*) operational improvements in the water and sanitation company (SABESP); and (*d*) studies. The first of three stages of the project was launched in 1992. The second stage, which started in 2002, will be concluded in 2007, when 350 million liters of wastewater are expected to be treated.

Urban Flooding

The new concept for a macrodrainage plan that integrates both municipal and state plans and programs, and proposes technical, economic, and environmental solutions, is one of the most important contributions of the Alto Tietê Committee. However, the elaboration of this plan, under preparation since 1998, has undergone several periods of paralysis. The studies that have thus far been carried out propose a series of structural and nonstructural interventions in the order of R$1 000 million (US$465.1 million). Another project of more or less the same size, proposed by an earlier plan, is already under way with funding from the Japanese government.

Sewage Collection and Treatment

The massive investments in treatment and collection began in the early 1990s, before the institution of a new water management system in São Paulo. SABESP estimates that the sewage gap will decrease substantially in coming years, falling from 17 percent in 2005 to 10 percent in 2010, 8 percent in 2015, and 7 percent in 2020. Even so, approximately 1.4 million inhabitants will remain unattended in 2020, a substantial number in absolute terms. There are no estimates for increases in wastewater treatment (FUSP 2002b).

Water Quality and Environmental Concerns

As has already been noted, massive investments in sewage collection and treatment have not been enough to improve water quality in the strategic reservoirs in the Alto Tietê basin. The only exception is the Billings reservoir, where quality has been improving dramatically since 1992, when the reversal of the polluted water from the Pinheiros River into the reservoir by the electricity company was restricted. This was a victory for several movements that became involved in water reform with the objective of protecting the reservoir for supply purposes.

5.5
Summary and Conclusions

5.5.1
Review of Basin Management Arrangements

The Alto Tietê river basin brings up many interesting questions around the issue of integrated water resource management at the lowest appropriate level. The analytical framework developed for the overall research project suggests that the political and institutional conditions in São Paulo and the Alto Tietê basin should have been favorable to the development of integrated and participatory management. However, almost fifteen years of reform have not been sufficient to make the new water resource management system fully operational anywhere in the state, and the outcomes have been much less impressive than was expected.

The reasons for this have been outlined in this chapter. Implementation has taken the path of least resistance, advancing only in areas that have been less costly in political terms, such as creating regulatory bodies (the state council and the basin and subbasin committees), the elaboration of water resource plans, and the execution of a new water permit system. However, when it came down to more controversial issues such as water pricing, political will weakened in the face of resistance from various government actors and organized user groups, and no strong champion emerged within state government with the political clout and will to further advance the decentralization agenda.

As outlined earlier, charging for water is one of the key issues in making the Alto Tietê Committee more relevant and giving it more say in water investment and management decisions. As long as such decisions remain at the individual agency level (both state and municipal), decisionmaking remains fragmented and key policy instruments to curb water demand increases and pollution are not implemented and used.

Water charging has only been implemented in Brazil where a strong and determined public actor mobilized to overcome the skepticism and active opposition of both government agencies and water users (Formiga Johnsson et al. 2003).[22] Those advances that have occurred in the Alto Tietê basin are largely a result of the great enthusiasm and commitment of the state técnicos who fought for the new water management model. Early in the reform process, the técnicos were able to obtain powerful allies who helped convince the state governor to buy into their ideas. The reform's promoters were able to persuade the Inter-American Development Bank to consider the law's enactment to be a precondition for approving the 1991 Tietê Project to clean up the rivers and reservoirs of the São Paulo area (Abers and Keck 2005).

The lack of government commitment to the process is not enough, however, to explain the lackluster performance of the Alto Tietê Committee. Several peculiarities of the Alto Tietê context made it even more difficult than elsewhere in São Paulo state for river basin bodies to take advantage of their favorable conditions and take the lead in coordinating water management. First, the extent and intensity of water-related problems (and solutions), typical of highly dense and industrialized regions, represent an enormous technical, political, and financial challenge. Under these conditions, it is harder for stakeholders to identify common interests. Second, the peculiar composition of the Alto Tietê Committee, which included among its members powerful state government agencies and the government of São Paulo municipality, has so far proven to be more of a problem than an advantage. These influential institutions have not needed to take the committee seriously thus far and it is unlikely that they will throw their energies into committee activities until the pricing system is implemented, as occurred with many powerful stakeholders in the Paraíba do Sul basin.

---

[22] Only four experiences are under way in the country: charges for state water in Ceará since 1996; charges for federal water in the Paraíba do Sul basin (Rio de Janeiro, Minas Gerais, and São Paulo states) since March 2003; charges for state water in Rio de Janeiro since March 2004; and charges for federal water in the Piracicaba-Capivari-Jundiaí basins (São Paulo and Minas Gerais) since January 2006.

Both the intensity of problems and the lack of mobilization of crucial committee members seem, however, to lose importance at lower levels of management. As forums for elaborating and implementing the water source protection policy at the local level (among other attributions), the subcommittees serve as strong building blocks for integrated management in the basin. Indeed, the lowest appropriate level for many water management functions turned out to be even smaller than the original division of the Tietê river basin into five regions. The subregions that were created in the Alto Tietê basin can be defined as "social catchment" areas, combining socioeconomic and environmental interests and identities with the region's political and natural hydrological divisions (Kemper 1998).[23]

Another conclusion of this chapter is that important achievements have been made, though the decentralization process has yet to reveal measurable physical results such as the improvement of water quality or the rationalization of water use. It is undeniable that the Alto Tietê Committee and its subcommittees have already played an important leadership role around several issues. Above all, an extraordinary mobilization around water issues, problems, and management has occurred, even though solving many water-related problems may be beyond the capacity of the committees or even of the water resource management system as a whole.

Finally, it should be stressed that the decentralization model, first developed by São Paulo and later confirmed in the federal legislation and most state laws, is well adapted to the conditions of the Alto Tietê basin. It is there, in the Brazilian state that has the richest, best qualified, and most experienced water management institutions, that the model centered around the river basin committees and basin agencies, with financial sustainability guaranteed through bulk water pricing, has the best conditions to be successful. However, implementing this model has proved slow, arduous, and generally challenging, to the extent that the pioneer state in water reform has begun to lag behind others.

5.5.2
Future Prospects

The above discussion has made it clear that the Alto Tietê basin still needs to advance in the clear definition of roles and relationships among the various organizations involved in river basin management. If both the basin agency and DAEE must act as executive secretary during this transitory phase, then their activities should at least be better integrated. The basin agency and the main committee (which is, officially, supposed to have authority over the agency) should coordinate discussion on the main problems and solutions in the basin. In this sense, the final agency's sphere of influence remains unclear. Will it be only the executive arm of the main committee or of all committees in the basin? The subcommittees also need to coordinate better with

---

[23] Kemper defines "social catchment" as a management unit within the larger hydrological basin, with common economic and social concerns; the social catchment concept permits the interests of local stakeholders to be taken into account and relates their interests and incentives to the natural environment.

the main committee in an integrated management system to be created at basin level. The main challenge is to transform the Alto Tietê Committee from a social force only into an authoritative arena for decisionmaking. This challenge is even more important in the current context, in which the dynamism of the main committee has declined.

Finally, in order for the water management system to be effective, a bulk water pricing system must be implemented. The importance of pricing is twofold. First, it will be fundamental for promoting the rational use and sustainability of water resources in the basin, principally by reducing the imbalance between water demand and availability that has characterized the Alto Tietê for decades. Second, by providing basin institutions with financial autonomy, the pricing system will make the committees more viable. Since it is likely that charge revenues will be accompanied by high-value investments, they will also contribute to building integrated management, with the participation of the major users and agencies who have until now paid little attention to the committee in the design and implementation of their own investments. However, as previously noted, this issue extrapolates to the river basin level. It is a challenge at the state level, perhaps the most important one, which is starting to be faced with the recent approval of the state Water Charges Law of 2006. Many hope that the effective implementation of this law will not take as long as its approval.

IBRD 33728

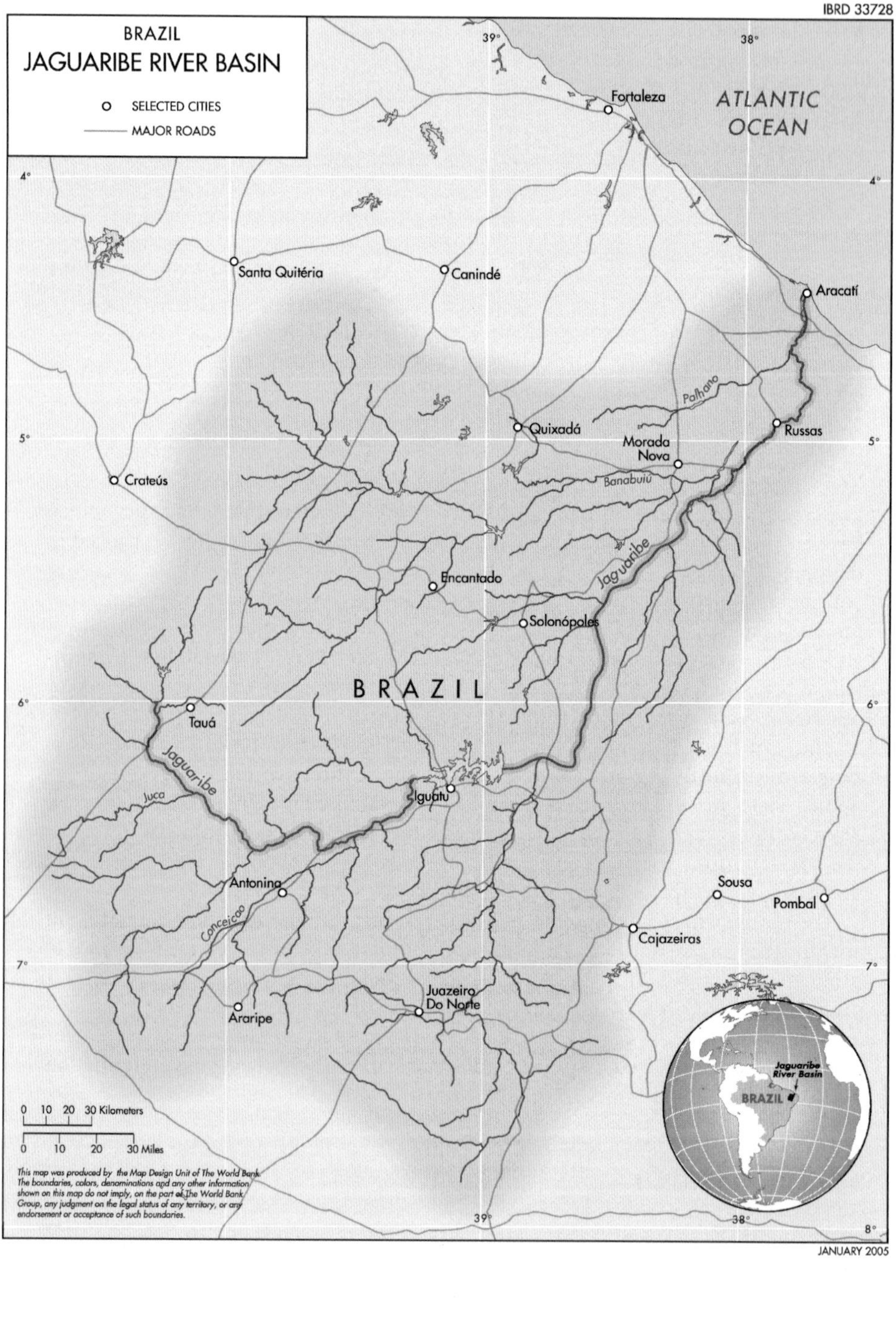

JANUARY 2005

Chapter 6

Brazil: Jaguaribe Basin

R. M. Formiga Johnsson · K. E. Kemper

6.1
Background

6.1.1
Introduction

Reform of water resource management in the Jaguaribe basin, Ceará state, occurred as part of Brazil's restructuring of its water resource management system, which has been ongoing since the early 1990s. New water legislation was first approved in the state of São Paulo (see Chap. 5) at the start of the 1990s, followed by Ceará in 1992 and subsequently by several other states. The national Water Law of 1997 confirmed the new system, which focused on the river basin as the territorial unit for planning and management, with decisionmaking placed in the hands of stakeholder committees and basin agencies acting as their executive arms. Bulk water charges would bestow financial autonomy upon the new institutions.

The Jaguaribe River flows roughly northwards to the Atlantic Ocean through the semiarid Northeast region of Brazil; its basin lies entirely within Ceará, one of the poorest states in the country. The occurrence of periodic droughts and a pronounced dry season prompted supply-side solutions to water shortage problems based on the construction of a large number of reservoirs, but this system was proving increasingly inadequate to deal with supply shortfalls and conflicts.

The pre-reform socioeconomic, political, and institutional structure did not seem encouraging for decentralized water management as a means of resolving these problems. But, based on a more centralized model of water management than that proposed by the national Water Law, Ceará state has involved large numbers of stakeholders in key water management questions and created a state water management agency and decentralized institutions that are impressively strong, considering the context within which they grew.

6.1.2
Basin Characteristics

The Jaguaribe basin is an independent basin in the Atlantic hydrographic region of Brazil's Northeast. It has a drainage area of 72 560 square kilometers, covering approximately 48 percent of Ceará state. The principal river runs from south to north for about 610 kilometers, flowing into the Atlantic Ocean (COGERH-Engesoft 1999d). The basin has 80 municipalities and more than 2 million people, about a third of Ceará's population. After an intense process of urbanization in recent decades, the majority of the basin's population (over 55 percent) now lives in urban areas, still well below the state and national averages (of 72 percent and 81 percent respectively). For management purposes, the Jaguaribe basin has been divided into five hydrographic regions: Upper Jaguaribe, Middle Jaguaribe, Lower Jaguaribe, and two subbasins, Salgado and Banabuiú (Fig. 6.1).

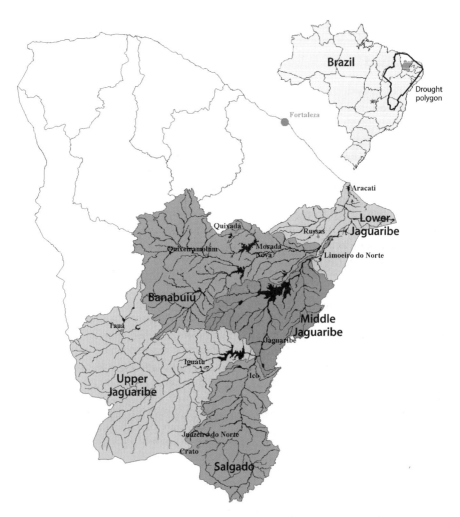

Fig. 6.1. Jaguaribe river basin in Ceará state: hydrographic regions and major cities (adapted from COGERH n.d.)

Most of the Jaguaribe basin falls within the semiarid region known as the Sertão (hinterland). Precipitation is highly variable, ranging from 400 millimeters in the hinterland to 1 200 millimeters along the coast. Although such rates of rainfall are higher than in many dry regions in the world, in Ceará the combination of impermeable crystalline rocks in the soil and high temperatures produce elevated rates of evapotranspiration – over 2 000 millimeters for the basin – and low levels of water retention and storage. Groundwater resources are considered of limited importance in most areas of the basin. Cyclical droughts occur at least every five years and can persist over a period of several years (COGERH/Engesoft 1999a–c).

The basin's rivers are naturally intermittent, flowing only during the rainy season. The state's – and previously the federal government's – main policy strategy has therefore been to store water resources in reservoirs for the dry season, and water resource infrastructure in the basin was already well developed before the decentralization process began. The basin has an estimated 4 713 reservoirs and a total storage capacity of 13 560 million cubic meters. Seventy-five percent of this water availability is provided by three reservoirs[1] which have transformed about 470 kilometers of rivers in the middle and lower part of the basin (Jaguaribe and Banabuiú valleys) into perennial waterways, directly benefiting 19 municipalities and changing the economic and political profile of the region. These reservoirs are also the main water sources for Ceará's capital city, Fortaleza, and its metropolitan area, the state's largest urban and industrial center, which, importantly, lies outside the basin.

In general, the basin is considered poor, even for Ceará, a state that contributes only 1.8 percent to the nation's gross national product.[2] The Jaguaribe basin follows the state trend, with most of its income deriving from the service sector. Although agriculture accounts for only a small part of the basin's income, it is of great social importance since subsistence agriculture (dependent on rainfall) still employs most of the rural and poor basin population.

6.1.3
Water Resource Problems

Water Scarcity and Recurrent Droughts

Until the beginning of the reform process in the early 1990s, the issue of water scarcity, exacerbated by unpredictable rainfall and recurrent drought, was treated as essentially a supply problem to be resolved through the massive construction of reservoirs and related water infrastructure (COGERH/Engesoft 1999e; Kemper 1996). This approach, however, has not prevented water use conflicts, providing no incentives against wasteful water use or in favor of water reallocation, and *vazios hídricos*

---

[1] The Orós reservoir (1.94 billion cubic meters), the Banabuiú reservoir (1.6 billion cubic meters), and the newly completed Castanhão reservoir (6.7 billion cubic meters).
[2] The GNP of Brazil in 1999 was R$1556 billion (US$554.9 billion). Data from IBGE 2006.

("waterless spaces", or regions without water storage systems) continue to exist, especially in the upper basin.

Growing Urban Demand and Interbasin Transfers

Population growth and urbanization in the basin have resulted in a rise in demand for water by domestic and industrial sectors. These sectors have increasingly competed with the dominant water use in the basin, irrigation projects promoted by the federal and state governments in the 1970s. In addition, since 1992 the Jaguaribe basin has become the main source of water for the expanding Greater Fortaleza region.[3] The construction of a new canal, the Integration Axis, from Castanhão reservoir (Middle Jaguaribe subbasin) to Fortaleza may increase consumptive use of the basin's waters by Greater Fortaleza to an estimated 43 percent during drought periods (COGERH/Engesoft 1999d). This has generated protests from water users in the Jaguaribe basin, fearful that diversions to Fortaleza could come at the expense of their own water security.

Intrabasin Water Scarcity and Allocation

Conflict over intrabasin water allocation results in large part from the great variations in climatic conditions and historic management practices. The most frequent conflicts arise between the users that depend directly on reservoir waters and those located downstream, and among users in the valleys that have been rendered perennial through regulation, henceforth referred to as "regulated valleys"[4] (Garjulli et al. 2002). An innovative management practice referred to as "negotiated water allocation" among users has demonstrated tremendous potential for reducing and even resolving some of these conflicts.

Water Quality and Environmental Concerns

Water quality is another major concern in the Jaguaribe basin. One of the main sources of declining water quality is the lack of municipal wastewater collection and treatment. The majority of urban areas have expanded without adequate sanitation infrastructure, and untreated sewage is commonly released directly into rivers and other bodies of water. In addition, agricultural practices in the region have generally given little consideration to such externalities as the impacts of excessive agrochemical use. River monitoring by the State Environment Superintendancy (Superintendência Estadual do Meio Ambiente; SEMACE) has not been systematic and only the three major reservoirs have any kind of water quality control program. Available data in the Jaguaribe basin plan (COGERH/Engesoft 1999f) suggest that there is substantial variation in the

---

[3] In 1993, after a three-year drought and a threat of extreme water scarcity, the state government built an emergency canal (Canal do Trabalhador) that can divert up to 5 cubic meters per second from the Lower Jaguaribe subbasin to the system of large dams that supplies Greater Fortaleza. Also, the new Castanhão reservoir, the largest in the state with a capacity of 6.7 billion cubic meters, was built mostly to meet increasing urban and industrial demand from Fortaleza.
[4] From the Portuguese *vales perenizados*.

water quality of the main river, and also indicate a high concentration of fecal coliforms in the major reservoirs at the beginning of the rainy season (January and February).

Recurrent Floods

While some years are very dry, excessive rainfall can occur in other years. When rainfall is high in the Jaguaribe basin, as occurred in 2004, floods affect various cities, especially those located near the Jaguaribe River in the lower part of the basin. This problem has yet to be dealt with effectively, although some reservoirs, such as the new Castanhão dam, the largest in South America for ephemeral rivers, were planned to reduce flooding in the cities.

Over the last century the increasing water management problems encountered in this semiarid region have been addressed through a top-down structure and massive investment in water infrastructure. However, drought and floods continue to affect farmers, and supply is increasingly struggling to match population and economic growth. The new management approach discussed below attempts to address these problems.

6.2
Decentralization Process

6.2.1
Pre-reform Arrangements for Water Resource Management

Before the current reform began in the early 1990s, water resource policy and management in the Jaguaribe basin, in Ceará state, and in the semiarid region more generally was traditionally the territory of federal initiative. The main federal agency for drought prevention – the National Department of Drought Relief (Departamento Nacional de Obras contra as Secas; DNOCS) – was created in the 1910s. Since then, it has expended large amounts of federal money for massive construction of water storage infrastructure. In the last century, close to 7 000 reservoirs were constructed in Ceará; about 130 are considered "strategic" and serve multiple water uses (Garjulli 2001a). While the largest and most strategic reservoirs were built by DNOCS, the state government focused on building small ones, usually during drought crises. This supply-based approach was characterized by a centralized, rigid, and untransparent decisionmaking structure that favored the interests of large landowners and the irrigation projects financed by DNOCS.

6.2.2
Impetus for Reform

The wider process of democratization in Brazil in the 1980s, which established a political climate ready to embrace the water management principles recommended by major international charters and organizations in the 1990s, including the Dublin Statement of 1992 and the World Bank in 1993, has been described in Sect. 5.2.2.

At the state level, this process of democratization was characterized by a new generation of state government leaders who had a more modern vision of development compared to the landed elite who had dominated the state's policies for so long, and who favored the creation of innovative institutional arrangements for water resource management at state and basin levels. This process was also supported by the World Bank, which at the time was negotiating a loan with the government of Ceará and made it a condition that further World Bank support to hydraulic infrastructure could only be provided if institutional reforms, as already defined in the state law of 1992, would be put in place to ensure long-term sustainability.

6.2.3
Reform Process

The first steps toward institutional change in Ceará were the creation of the Secretariat for Water Resources (Secretaria dos Recursos Hídricos; SRH) and the approval of the state Water Law (11.996/1992). The law embraces the main ideas of modern water resource management, also followed by the later federal Water Law (9.433/1997) and other state laws: integrated water management, with the river basin as the planning unit; water as a finite and fragile resource and as an economic good; and decentralized and participatory management. Likewise, Ceará included the same management instruments later instituted by the federal law: state and basin water resource plans; bulk water use permits; bulk water charges; and a water resource information system.

However, the political and institutional organization proposed in the state Water Law is more centralized than in many Brazilian states and in the federal law. For instance, the basin committees have fewer deliberative powers for some issues, especially the definition of the bulk water pricing system, which elsewhere is one of their primary attributions. Also, in Ceará basin committees will not have their own executive support structures (water agencies); the state law defined that technical support should be provided by state management institutions.

While most states relied on existing environmental or water agencies funded through the general state budget, in Ceará a strong, independent and self-financed Water Resources Management Company (Companhia de Gestão dos Recursos Hídricos; COGERH) was created in 1993 to carry out management, monitoring, and enforcement functions and to eventually take over control of federal infrastructure in the state, until then mostly governed by DNOCS. COGERH has recently created seven regional offices, four of which are in the Jaguaribe basin.

COGERH had to struggle for control over state water resources with DNOCS, which had its headquarters in Fortaleza and the larger part of its infrastructure domain in Ceará, and was initially reluctant to cede responsibility to the state; and with the Ceará State Water and Sanitation Company (Companhia de Água e Esgoto do Estado do Ceará; CAGECE), which was receiving payment for water from Greater Fortaleza's industries and was unwilling to give up this source of revenue. An agreement was eventually reached that allowed COGERH to take over the metropolitan basin system, with both industries and CAGECE being charged for bulk water, and CAGECE continuing to supply Greater Fortaleza with water. However, CAGECE pays far less than industries for bulk water, as the state government has not raised tariffs for domestic water supply and sanitation services to levels that would allow CAGECE to be financially self-sustaining (Kemper and Olson 2000).

The aim has been to centralize the technical aspects of water management as well as the collection of water charges in COGERH, with the objective of financing both its administrative expenses and the operation and maintenance of the water infrastructure for which it is responsible. The decision to centralize water charging is based on the need to redistribute resources among basins in the state, since – except for the Greater Fortaleza basin – none could be expected to cover their own operating and maintenance expenses.

The creation of basin institutions has occurred gradually over more than ten years, under the initiative and coordination of COGERH and with the support of SRH. The result has been the emergence of various types of local organizations, with different features, attributions and territorial scales of management, which partially overlap (Fig. 6.2):

- The Jaguaribe-Banabuiú user commission, which basically defines the annual operating rules of the three major reservoirs of the basin, according to the negotiated water allocation between the users of the regulated valleys.
- 36 user commissions of "isolated strategic reservoirs", i.e. those that guarantee multiple water use in locally important reservoirs during drought periods.
- 5 subbasin committees, corresponding to three parts of the basin (Upper, Middle, and Lower Jaguaribe) and to two basins of Jaguaribe's tributaries (Salgado and Banabuiú subbasins). Together, they cover the entire territory of the Jaguaribe basin.

The creation of user commissions was based on the realization, highlighted by the conflicts that arose during the major 1992–1994 drought, that focusing solely on hydrographic regions (basin, subbasin or part of a basin), as called for in the Water Law, was not the best approach in the semiarid Northeast, where local stakeholder interactions were most intense around reservoirs and along the regulated river valleys (Oliveira et al. 2001). The main purpose of the reservoir user commissions was to guarantee multiple water use in the immediate surroundings of the reservoir during drought periods when rivers dry up, with users and other local stakeholders, in a transparent process, deciding the volumes to be released from the reservoirs as well as the use and conservation rules that must be respected by all users. The result has been a substantial reduction in water use in the Jaguaribe basin, even under rationing conditions (COGERH 2000–2002). However, the user commissions are still only informal institutions.

The negotiated allocation of water through the user commissions coexists uncomfortably with an emerging formal water permit system. Water use permits, which the state SRH has slowly begun to issue with the help of COGERH (COGERH/Engesoft 1999g; Baltar et al. 2003), are not required to respect the decisions of the user commissions. A proposal is on the table to grant permits for longer periods, coupling them with the negotiated allocation process as it is practiced, for example, in the Northern Colorado Water Conservancy District, United States (Kemper 1999). This would give users the security of holding permits for a specific period while specific amounts would be negotiated each year, based on water availability.

The five subbasin committees that exist today in the Jaguaribe basin were created only several years after stakeholder participation was established through the

Fig. 6.2. Decentralized institutions of the Jaguaribe basin (developed based on COGERH and field data)

commissions (Garjulli et al. 2002). The committees have broader water management attributions than the commissions, such as setting guidelines, approving basin plans, and conflict resolution. Their creation, which occurred between 1998 and 2001, was a much more formalized process that had to comply with both national and state regulations.

Over time, considerable changes have thus been made in the institutional arrangements originally defined by the state Water Law, though it can be said that the spirit of the law has not been lost sight of. The resilience of the changes was demonstrated by the renewal in 2004 of the agreement between DNOCS and Ceará's government through which the former shared powers with COGERH over the management, operation, and maintenance of federal reservoirs.

Indeed, Ceará is the state that has progressed the furthest in terms of implementing water reform. It was the first state – and the only one until 2003 – to implement a system of bulk water charges, which are currently levied on domestic, industrial, and some irrigation water uses. This has given financial self-sustainability to the agency. Water management and allocation decisionmaking for strategic reservoirs has become more democratic and participatory, evolving into a sort of informal water rights system. The state and most of the river basins now have water resource plans that reflect more comprehensive and higher-quality knowledge about water problems.

Problems remain. For example, the introduction of water charges is meeting opposition in the Jaguaribe basin. Given that water in the basin is mostly used for low-value agriculture, irrigators are largely unwilling and unable to pay at levels that would compensate for what COGERH spends on operating and maintaining water infrastructure in the basin.[5] An attempt in 2001–2002 (the Águas do Vale (Valley Waters) program) to introduce charges for water for irrigation, combined with a strategy to encourage rice producers to shift to less thirsty and more profitable crops such as cantaloupe and banana, was successful in that water consumption was reduced, but efforts to charge those who did not change their practices collected only about 20 percent of the amount charged (da Silva 2003).

COGERH has thus followed another strategy: to start by charging the major users in the metropolitan basin, which now contributes over 90 percent of the total collected revenues. In 2005, COGERH did start a large expansion of the state water charges system, and is gradually including irrigation, shrimp farming, fishing, and other uses, though it is facing criticism from water users and civil society organizations that it is being nontransparent with respect to how it defines prices and determines how proceeds will be used. Local stakeholders apparently do not, however, question the centralization of pricing at the state level; most users are served by public water supply infrastructure and most of Ceará's basins are underdeveloped and therefore benefit from the transfer of revenues from charges from the metropolitan basin.[6]

Institution of the bulk water pricing system is being met with substantial resistance from users and civil society at large because they fear the creation of a water market in the Jaguaribe basin. Even COGERH and SRH officials are generally unconvinced about the idea of water trading, even though a pilot water market project had been agreed between Ceará state and the World Bank.

---

[5] For an exhaustive analysis of this issue in the Curu basin, see Kemper 1996.
[6] In most of Brazil, basin committees not only negotiate the values of the charges directly with users but also decide how proceeds should be used, according to the priorities set by basin plans. This is what is proposed by the federal and most of the states' water legislations, and applied since 2003 in the Paraíba do Sul basin, in the southeast of the region. More details can be found in Formiga-Johnsson and Lopes 2003.

6.2.4
Current Situation

Table 6.1 indicates the main institutions for water resource management operating in the Jaguaribe basin, and their principal functions. COGERH, the state water manage-

Table 6.1. Institutions for water resource management in the Jaguaribe basin

Management level	Institution	Current water management attributions *Financial aspects*
Federal government *Water legislation, hydropower, and management and control of federal waters*	ANA National Water Agency	Establishment of the national water resource policy and system. Priority for combating pollution and drought. *Administrative and financial autonomy, funded by the federal budget and by royalties from the hydropower sector*
	DNOCS National Department of Drought Relief	Protection against drought and flooding, promotion of irrigation in semiarid regions. In Ceará state: management, operation, and maintenance of reservoirs and infrastructure built with federal money (nowadays those considered strategic for integrated water resource management are primarily run by COGERH). *Funded through the federal budget*
State government *Management and control of state waters*	SRH Secretariat for Water Resources	Establishment of state water resource policy. Coordination of the state water management system. Responsible for issuing and controlling water permits, with the technical support of COGERH. *Funded through the state budget and international loans*
	COGERH Water Resources Management Company	Responsible for the planning and management of state waters, operation and maintenance of the hydraulic system, water allocation, introduction and implementation of bulk water charge system, and organization of and interaction with user commissions and basin committees. *Funded through bulk water charges*
	FUNCEME Foundation for Meteorology and Water Resources	Responsible for meteorological monitoring in Ceará. Provides technical support to COGERH for simulations of reservoir operations and water allocation. *Funded through the state budget*
	SOHIDRA Hydrological Works Superintendency	Responsible for building and maintaining the state's water resource infrastructure. *Funded through the state budget and international loans*
	CAGECE Ceará State Water and Sanitation Company	Provides urban water supply to Greater Fortaleza and most municipalities in the interior of Ceará. Provides sanitation services only to Greater Fortaleza and a few municipalities in the interior. *Funded through water fees and the state budget. Limited recovery of operation and maintenance costs*

Table 6.1. *Continued*

Management level	Institution	Current water management attributions *Financial aspects*
River basin	Subbasin committees *Formal institutions under state jurisdiction and regulation*	Correspond to the Upper, Middle and Lower Jaguaribe, and to the Salgado and Banabuiú subbasins. *No systematic funding. Occasional financial support from COGERH*
	Jaguaribe and Banabuiú Valleys Commission *Informal institution created by the state (COGERH)*	Covering the regulated parts of the Jaguaribe and Banabuiú valleys (see Fig. 6.2), it includes water users, civil society, and representatives of key institutions. Decides the annual operation rules for the three major reservoirs of the basin through negotiations among its members about water allocation, and under the leadership and guidance of COGERH. *No systematic funding*
	Reservoir user commissions *Informal institutions created by the state (COGERH)*	The 36 reservoir commissions include users and other stakeholders interested in or affected by water allocations in the area of hydrological influence of each "isolated strategic reservoir". Allocation process similar to Jaguaribe-Banabuiú Valleys Commission. *No systematic funding*
Municipal government *Land use, urban drainage, and water supply and sanitation*	Municipal secretariats related to urban water infrastructure and management issues	Responsible for managing land use and occupation. Share responsibility with state government for urban drainage and local environmental issues. *Funded by municipal budget*
	Municipal water and sanitation services	Only a few municipalities are not supplied by CAGECE and have their own local urban water supply services. *Funded by water fees and municipal budgets. Limited recovery of operation and maintenance costs. No financial or technical assistance from the state. Partially supported by the Federal Health Agency (Fundação Nacional de Saúde; FNS)*

ment agency, has certainly become the most pivotal water management institution in the state, involved in all aspects of water policy. It has taken on responsibilities previously under other jurisdictions, including control of infrastructure, and has managed to establish a working relationship with both DNOCS and CAGECE (Kemper and Olson 2000). It has also created new organizations and policy mechanisms, and has demonstrated its ability to find innovative solutions to water management problems, as with the creation of a flexible system of user committees and subbasin committees.

6.3
Application of Analytical Framework

Examining the factors identified in the analytical framework (see Chap. 1) will contribute to our ability to understand the impact of decentralization on the Jaguaribe basin and on integrated water resource management.

6.3.1
Contextual Factors and Initial Conditions

At the time that reform began, local conditions in the Jaguaribe basin seemed unfavorable to the development of a decentralized and integrated water resource management system in several respects.

First, the basin is characterized by poverty, even in the Brazilian context. The political and economic development that occurred in the middle and lower parts of the basin (Jaguaribe and Banabuiú valleys) during the 1970s – a result of the construction of very large reservoirs – was not enough to change this general picture. Second, the proposal of participatory water management diverges in fundamental ways from the political culture of the Ceará Sertão, where water had historically been considered either a private good, the property of the owners of the lands through which it flowed, or under the control of the government agencies of the reservoirs within which it lay (usually DNOCS). Third, the dominant usage of water resources by irrigators strongly favored irrigation projects promoted by the federal and state governments in the 1970s. Irrigators in the basin account for about 83 percent – or 342 million cubic meters per year – of total water consumption in the basin, with 37 percent for private activities and 46 percent for government projects. Most of the conflicts in the basin involve this sector.

At the same time, however, other factors favored reform, especially in the broader context within which the changes would take place. At the national level, democratization of politics in the 1980s encouraged decentralization and participation in policymaking. The technical water resource community, led in large part by the Brazilian Water Resources Association, began to promote integrated water resource management models including water use rights, pricing, and basin-level management (Formiga Johnsson 1998). The impact of these developments was felt at state level, with a new more forward-looking generation of state leadership in Ceará encouraging the creation of innovative institutional arrangements for water resource management at state and basin levels.

6.3.2
Characteristics of Decentralization Process

The decentralization process in the Jaguaribe basin was marked by two distinct phases: (a) decentralization from federal to state level, a result of the increased technical, institutional, and financial capacity of Ceará's water resource management agencies; and (b) decentralization from state to local level, which occurred through the creation of deliberative and consultative bodies at the river basin and lower territorial levels.

The creation of COGERH was not part of the original design called for in the state Water Law, but resulted from the World Bank's insistence that the state create a water agency with management, monitoring, and enforcement functions, including pricing and the involvement and organization of local stakeholders (Kemper and Olson 2000). The fact that in 1997 COGERH took over some of DNOCS's management responsibilities represented a major step towards decentralization from federal to state level. The sustainability of this federal to state decentralization process was demonstrated by the renewal of the contract between DNOCS and COGERH in 2004. Indeed, there are signs that federal-to-state devolution is increasing, such as a recent agreement between COGERH and the National Water Agency (Agência Nacional de Águas; ANA) delegating authority for issuing user permits for waters in federal reservoirs.[7]

The state-to-local decentralization process has been more complex. It can clearly be characterized as a top-down initiative, in which COGERH's User Mobilization Department (Departamento de Organização de Usuários; DOU) played a central role in managing a delicate balance between the sometimes centralizing tendencies and interests of a state-level institution, and local interests and customs. However, conservative elements in state government, and in the upper echelons of COGERH, remain distrustful of the participatory decisionmaking bodies that have been created, and in 2003 the DOU was dismantled. Although the long-term impact of these changes on the user commissions and basin committees is still unclear, reform-oriented officials hope that the high level of mobilization achieved in the basin over the last ten years will make it difficult to undo the advances made thus far (Lemos and Oliveira 2004).

6.3.3
Central-Local Relationships and Capacities

The devolution of some of DNOCS's authority over the management and control of reservoirs to Ceará state has been highly effective, since COGERH has developed substantial technical, administrative, and financial management capacities. Currently, COGERH operates and manages, through its agreement with DNOCS, all major reservoirs in the state, accounting for over 90 percent of the state's water storage.

But other aspects of water management remain underdeveloped. The state has proceeded only slowly with the implementation of groundwater management. The development of a new water use permit system has also been slow, despite the fact that criteria and procedures were basically defined some time ago. A recent World Bank study characterized the permit system as still in consolidation (Baltar et al. 2003).

Conversely, decentralization in the Jaguaribe basin has gone furthest with the user commissions, especially through the negotiated allocation of water, which, as discussed above, has proven very effective. However, within COGERH and SRH (to which COGERH is subordinate) there has been resistance to giving decentralized bodies greater power over water management. The result is that only the subbasin committees have been legally created, but these have received little real support or authority. Meanwhile the Jaguaribe-Banabuiú Valleys Commission – where the process of participatory decisionmaking began and has continued with great intensity – is still only an informal institution. This contradictory situation has created tensions between the subbasin committees and the Jaguaribe-Banabuiú Valleys Commission (see Sect. 6.3.4).

---

[7] The fact that only one of Ceará's rivers is federal could lead one to believe that the relationship between federal and state government is less relevant in Ceará. This interpretation would be, however, incorrect, since the constitutional norm of 1988 (Art. 26) grants federal control over waters collected by federal projects, even when these are built on state rivers.

6.3.4
Basin-Level Institutional Arrangements

Internal basin-level institutional arrangements are currently the source of some controversy. Some subbasin committee executive board members have argued that the Jaguaribe-Banabuiú Valleys Commission should be dismantled, with the transfer of its responsibilities to the subbasin committees. During 2005, this movement has grown stronger and it seems that the committees are taking over the water allocation responsibilities. However, before rethinking the attributions of these local bodies and their relations with each other, it is necessary to determine the extent to which decentralization down to units that are smaller than the river basin has been positive.

The division of the Jaguaribe into three parts (Upper, Middle, and Lower) was based not on hydrological, social, or economic criteria but on logistical and operational criteria, since COGERH's DOU was unable to operate immediately in the entire basin. The solution was to create committees in subbasins in the short run, with the plan to join them together later into a single committee. Some still support this plan, while others have proposed creating a separate basinwide committee whose objective would be to coordinate decisions made by the smaller-scale bodies.

There is, however, general support for the 36 reservoir user commissions. The allocation process they engage in is similar to that carried out by the Jaguaribe-Banabuiú Valleys Commission, but the decisions have only very localized impact and transaction costs are lower. Usually, the commissions include only users or groups of users directly affected by water allocations in the area of hydrological influence of a single reservoir, since members are mostly made up of organizations working in the perimeter of the reservoirs, and in the immediate downstream area.

6.4
Performance Assessment

6.4.1
Stakeholder Involvement

The creation of subbasin committees and user commissions, under COGERH's coordination with the support of SRH, has allowed for the involvement of hundreds of stakeholders of all types, such as municipalities, public and large private irrigators, fishermen, and industry leaders. Although so far stakeholder involvement has been limited largely to the negotiated allocation of water and to conflict resolution, these experiences are still a radical transformation in management practices. Also, the participatory nature of the process appears to increase users' sense of ownership, that is, they are not only users, but also managers and "stewards" of the resource.

However, local stakeholders still have no say in some decisionmaking processes that affect them directly, such as bulk water pricing or interbasin transfers to Greater Fortaleza, which continue solely under the control of state government agencies. In this sense, the decentralization proposed in the state Water Law was not translated into practice. Also, the concerns of local stakeholders with respect to water quality problems and broader environmental problems related to water have yet to find a place on the agenda of the state water institutions.

Thus while federal-to-state decentralization has undoubtedly advanced substantially, the process of devolving authority and responsibility from the state to local levels is less easy to characterize in terms of success or failure. The basin committees are formal institutions that still have not found a de facto place in the water management system, and lack effective technical, administrative, and financial support to carry out their attributions. Also, as Ceará has centralized water charging at the state level, basin committees will not have their own executive arm (basin agencies), nor will they have financial resources of their own, and their activities have been limited to information dissemination, consciousness raising, capacity building among local actors, and the resolution of water use conflicts.

Water allocation within the basin has improved considerably. Traditionally, certain user groups had priority in gaining access to water, especially irrigation projects run by DNOCS, large users and agribusiness, and, of course, human consumption, which is guaranteed first priority by law. Generally, these privileged users are also the most organized groups in the basin, with greater economic and political resources. Those who have not been privileged historically – such as fishermen, *vazanteiros*,[8] and smaller irrigators – would however theoretically benefit most from formal water rights, decentralization, and participatory management (Kemper 1996). And, in practice, such groups have indeed benefited most over the last decade from the "negotiated allocation" processes, through which groups representing larger numbers of less powerful users have finally gained a voice.[9]

It remains to be seen to what extent water users will be involved in decisions related to such management instruments as bulk water charges, water permits, and allocation mechanisms. Water users want to have more of a say in these – guarding their interests to reliably receive water at the lowest possible cost to them – while the state seems to assume that the involvement of users at the current level, that is mainly discussions, information sharing, and negotiated allocation, is sufficient devolution of power to the basin level.[10]

Despite the current uncertainties concerning institutional boundaries, both user commissions and basin committees have been promoting the resolution of water use conflicts, with the support of COGERH. They have come to be perceived as the legitimate space for negotiating conflicts over water allocation and quantitative use, and for airing other controversial issues related to water quality and environmental degradation.

6.4.2
Developing Institutions for Integrated Water Resource Management

As has been demonstrated, the devolution of federal authority over the management and operation of federal reservoirs to Ceará state has been highly effective.

---

[8] When the water level in a reservoir recedes, due to outflows or evaporation, the rim of humid soil that is uncovered is called a *vazante*. Farmers, mostly small subsistence farmers, who cultivate in these areas are called *vazanteiros*. They can also be found along rivers.
[9] These conclusions are based on interviews in February 2004, confirmed by Ballestero (2004), cited by Lemos and Oliveira (2004).
[10] Interviews during field visit in July 2004.

Federal institutions continue to develop, support, and finance specific drought relief programs in the semiarid region, together with state governments and sometimes with international organizations. The Águas do Vale program (see Sect. 6.2.3), which can be considered a demand-supply approach for irrigation, is an example of such partnership.

Decentralization from state to local level has been more partial. Although COGERH has decentralized the allocation of strategic reservoir waters to local institutions, many traditional water management attributions continue under COGERH's purview, such as water permit concession, bulk water pricing, planning, operation and maintenance of hydraulic infrastructure, and groundwater management and control. Furthermore, none of the changes introduced thus far have affected municipalities, which are still fully responsible for land use and urban drainage.

6.4.3
Effectiveness and Sustainability

At the basin level, the financial resources of decentralized institutions are both precarious and insecure. Neither the basin committees nor user commissions have their own financial resources, depending totally on contributions from the state government and from their own members. This makes them vulnerable to any top-down changes that may occur.

At the state level, though, bulk water pricing has represented an important change in terms of financing water management. Until pricing was introduced, bulk water supply services were partially or fully subsidized by public institutions. The pricing system has enabled COGERH to gradually achieve financial sustainability for its operation and maintenance costs and for investments in new water infrastructure (Azevedo and Asad 2000). Although Ceará is one of Brazil's poorest states, collected revenues, statewide, have grown substantially over time, from R$268 410 (US$124 850) in 1996 to about R$23 million (US$10.7 million) in 2005. With the current expansion, COGERH expects to collect R$29 million (US$13.49 million) in 2006.[11]

The metropolitan basin – the state's principal urban and industrial area – contributes over 90 percent of the total revenues from bulk water pricing. Among the user-payers, the domestic sector is currently the largest contributor (65 percent of the total), followed by industry (34 percent). Irrigation contributed only 1 percent of the total collected revenues. The result is that the degree of cross-subsidization within the state for water management costs is enormous, both among user sectors (from industries to the domestic and irrigation sectors) and regionally (from the metropolitan basin to the other basins). This means that the operation and maintenance costs of the large water infrastructure in the Jaguaribe basin are currently subsidized by users in the metropolitan basin.

It is likely that even with the expansion of the bulk water pricing system, expected to begin in 2005, proceeds generated within the basin will still be insufficient for the investments in infrastructure and water monitoring that are planned. In effect, most

---

[11] 1 US dollar = 2.15 Brazilian reals (March 2006).

of the water basin's users are irrigators who pay almost symbolic amounts. Therefore, we can expect both continued subsidization from the metropolitan region and government funding for larger infrastructure projects to be necessary.

The water allocation process has become more effective, efficient, and equitable compared to the process followed by DNOCS in the decades before COGERH took over. Nowadays water users are convened every year after the rainy season and informed about water availability, including stochastic model results for the coming year. This is the foundation of the negotiated allocation process, which permits water users to plan their production accordingly once the shares of each one have been agreed.

The process is not only more efficient but also more equitable, because traditionally weak user groups are included, get access to information, and have a kind of informal water right. This system considerably reduced the moral hazard approach in which DNOCS would keep information to itself and supply water users – usually the well-connected ones – to its liking. The negotiated water sharing system also permits water users to avoid the impacts of dry years and thus become more drought resistant.

Water management in the Jaguaribe basin, and Ceará more generally, has, thus far, focused on improving water infrastructure and optimizing use and allocation, the privileged arenas of hydrological engineering. Broadening the scope of basin management to include, for example, water quality management, ecosystem preservation, and other environmental issues has yet to come. Despite these concerns, it seems that the most pressing agenda for stakeholders in the basin is not to expand the scope of the water resource management system, but rather to consolidate the advances made thus far.

6.5
Summary and Conclusions

6.5.1
Review of Basin Management Arrangements

The case study of the Jaguaribe basin is a fascinating example of integrated water resource management at the lowest appropriate level. In the first place, it suggests that even when preexisting conditions are almost entirely unfavorable, changes leading to more integrated and decentralized practices are possible. The analytical framework developed for the larger research project within which this case study is located (Chap. 1) would suggest that the political and institutional situation of Ceará and the Jaguaribe basin was largely adverse to the increase of stakeholder involvement and transparency in decisionmaking. But when the decentralization model was tailored to these conditions, it was possible to begin to overcome them.

A second issue is how these changes occur. What happened in the 1990s that made it possible to transform practices that had been operating for decades? Water scarcity and conditions of almost permanent rationing certainly were motivations for change. But these conditions existed before. Perhaps more important was, on the one hand, the postdictatorship national context, which was highly favorable to democratization and decentralization; and on the other hand, a movement for reform specifically within the water resource sector had been disseminating a culture of integrated, participatory, and economically sustainable management throughout the country since the 1980s. Both of these conditions, however, could describe all Brazilian states, most of which did not make the advances in decentralized water management that we have described in Ceará. Certain conditions specific to the Ceará context allowed that state to take advantage of these conditions where others could not. A combination of an innovative state government, with an entrepreneurial orientation, and strong support from the World Bank for reform in the water sector were critical for the fact that water security and management made it to the top of that state's political agenda.

The high incidence of poverty in Ceará, its regional disparities, the limited capacity of user sectors to pay for water, and the high cost of bulk water supply have, however, been the reason why the Ceará law does not entirely correspond to the decentralization model that was later developed in the federal legislation and most state laws. That model is centered around the creation of basin committees and basin agencies with financial sustainability guaranteed through bulk water pricing. But in the Jaguaribe basin – and in Ceará as a whole – the state government began to play a much more proactive role in water resource management, primarily through COGERH. In a sense, the adaptation made in the Ceará case was simply less decentralization from the state to local levels than the national model and even the state's own Water Law proposed. The presence of large hydraulic structures throughout the state, which must be operated in close coordination if recurrent droughts are to be dealt with effectively, justifies this more centralized system. At the same time, what is particularly interesting about this approach is that although it is more centralized, local mobilization and stakeholder involvement is more intense than anywhere else in Brazil.

A third conclusion of this study is that the lowest appropriate level for decentralization is not always the river basin. As forums for negotiated allocation and conflict resolution, the user commissions serve as strong building blocks for integrated management. The subbasin committees are still trying to define their roles and powers. Their creation, however, is a consensus at local level and they have increasingly mobilized local actors around water issues. The essence of Ceará's experience in the Jaguaribe basin may thus be that the basin scale is less relevant there for integrated water management purposes, in favor of combining state-level management with decisionmaking at smaller territorial levels than the basin, such as subbasins, regulated river valleys, and reservoirs.

Finally, it should be stressed that much remains to be done, especially with respect to building a more holistic management system that incorporates efforts to promote better water quality and to coordinate water and environmental management. Nonetheless, the achievements made thus far are remarkable when compared to the problems and practices that seemed, until recently, impossible to overcome. Water rationing in the Jaguaribe basin used to be an almost permanent state of affairs. Traditional institutions used to privilege the interests of entrenched oligarchies. Civil society and small users were excluded from water-related decisionmaking. Water was, in general, managed and protected in only the most precarious and unsustainable of ways. All these unfavorable factors have strongly challenged and will continue to challenge efforts to build a decentralized and integrated water resource management system in the Jaguaribe basin. The achievements already made, however, are quite impressive.

6.5.2
Future Prospects

As regards future priorities, policymakers first have to establish how negotiated water allocation fits into the formal institutional arrangements for water resource management. Proposals for the combination of interannual permits (issued by the state) and annual negotiations to determine the available amount under rationing conditions (decided informally by the user commissions) seems to be a good way to adapt water allocation policy to the specific conditions of the semiarid region. This would eliminate the contradiction between a formal but weak system and an informal but highly legitimate one.

Second, the state government has yet to overcome the skepticism of the irrigation sector with respect to the bulk water pricing system. The expansion of the charging system, currently under way, has been criticized by local stakeholders, who complain that COGERH and the state government have made unilateral decisions, with little debate with basin committees.

Many of these problems are, however, the result of the fact that the state has failed to demonstrate clarity and certainty with respect to the level of decentralization to be pursued in the water management system. Indeed, COGERH has yet to define both its own role in basin organization and the relationship it should have with basin committees and user commissions. Some argue that if the system is to be made compatible with the spirit of the water laws, that is, if it is to become more decentralized and participatory, COGERH's current policies towards the basin-scale institutions would have to be reformulated substantially (Teixeira 2004, and personal communication with local representatives 2004). Above all, COGERH will have to decide if it will become a water agency for all the state's basins, instead of serving as only a sort of occasional support mechanism. If it chooses to take this path, it will need first to build legitimacy and this will require breaking with traditions of closed-door decisionmaking and centralized administration, especially with respect to bulk water pricing. It will then have to adopt mechanisms gradually leading to a system in which management responsibilities are shared with those who pay for bulk water use and in which the committees have deliberative power and can supervise its activities. But the recent demobilization of the DOU – which has functioned to date as the main liaison between COGERH and local organizations – sends a signal in the opposite direction.

The success that has made the Ceará model so famous in Brazil and internationally has depended on two factors: the capacities and expertise of COGERH and the involvement and mobilization of local stakeholders. But not only are COGERH and local stakeholders in constant tension and even competition, but also the various stakeholder bodies that have been created have yet to establish a clear division of labor. Consolidating Ceará's success will depend on finding the right balance between these forces.

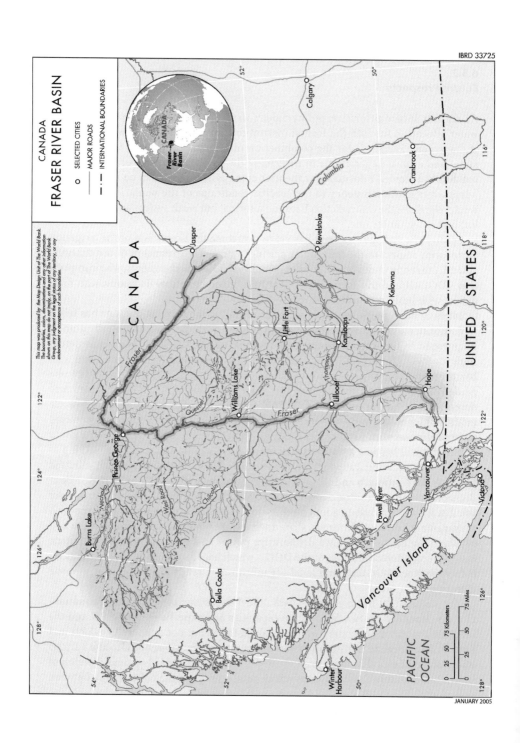

CANADA
FRASER RIVER BASIN

o SELECTED CITIES
— MAJOR ROADS
–·– INTERNATIONAL BOUNDARIES

This map was produced by the Map Design Unit of The World Bank.
The boundaries, colors, denominations and any other information
shown on this map do not imply, on the part of The World Bank
Group, any judgment on the legal status of any territory, or any
endorsement or acceptance of such boundaries.

CANADA

Jasper

Columbia

Calgary

Revelstoke

Cranbrook

Fraser

Little Fort

Kelowna

Williams Lake

Kamloops

Quesnel

Thompson

Fraser

Lillooet

Prince George

Nechako

West Road

Chilcotin

Hope

UNITED STATES

Burns Lake

Powell River

Vancouver

Bella Coola

Victoria

Vancouver Island

PACIFIC
OCEAN

Winter
Harbour

0 25 50 75 Kilometers
0 25 50 75 Miles

52°
50°
116°
118°
120°
122°
122°
126°
128°
54°
52°
50°
124°
126°
128°

Fraser
River
Basin

CANADA

JANUARY 2005

Chapter 7

Canada: Fraser Basin

W. Blomquist · K. S. Calbick · A. Dinar

7.1
Background

7.1.1
Introduction

Basin-scale institutions for water resource management have emerged relatively recently in the Fraser basin; the Fraser Basin Management Board was established only in the early 1990s, to be succeeded in 1997 by the Fraser Basin Council. The case is of interest within this series of studies for three reasons: first, it adds an example of nongovernmental river basin organization, whereas the other cases are of governmental or intergovernmental structures; second, the Fraser Basin Council has pursued a very broad scope of topics that its members see as related to an over-all concept of basin sustainability, which includes social and economic as well as environmental aspects; and third, the formation of the Fraser Basin Council (and its predecessor Basin Management Board) was a locally initiated action that occurred in the fairly recent memory of many individuals who are still actively involved, and whose perspectives on the origin and evolution of the basin management effort are both fresh and rich.

7.1.2
Basin Characteristics

The Fraser River drains 238 000 square kilometers of British Columbia, Canada, an area about the size of the United Kingdom. The Fraser river basin supports a population of more than 2.7 million residents, and an economy that includes forestry and pulp and paper production, ranching and agriculture, fishing, mining, recreation, and tourism. Seventy-eight percent of the basin's population lives in the lower Fraser valley and estuary region, where the Vancouver metropolitan area is located.

The basin has been home to aboriginal peoples, or "First Nations", for thousands of years. The current population of indigenous residents is estimated to be 50 000. The number of distinct First Nations is subject to varying estimates, but Fraser Basin Council estimates place it at around one hundred, which may be categorized into eight major language groups.

The river itself is 1 399 kilometers long, originating in the Rocky Mountains and emptying into the Strait of Georgia and the Pacific Ocean after flowing through the Vancouver metropolitan area. There are 13 principal watersheds or subbasins of the Fraser basin, identified in Fig. 7.1, but on a broader scale three main hydrologic

Fig. 7.1. Fraser River subbasins.
Source: Fraser Basin Council
*(www.fraserbasin.bc.ca/
fraser_basin/watersheds.html)*

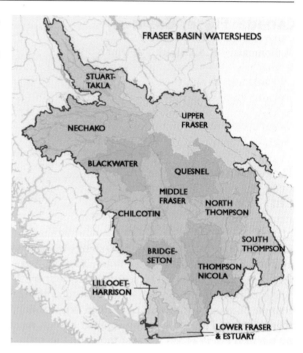

regions can be identified – the coastal mountains, the interior plateau, and the eastern (Columbia and Rocky) mountains. The interior plateau is the driest of these regions, the coastal mountains the wettest.

Weather systems moving onshore from the Pacific deposit large volumes of precipitation during autumn and winter in the mountain ranges, increasing with elevation and occurring predominantly as snow that thaws through the spring and summer months. While snow is also the principal winter precipitation in the interior plateau, peak precipitation in this region occurs as spring and summer rainfall. In some areas of the basin, glacial melt is also an important source of surface stream flow.

Between snowmelt from the mountains and rainfall in the interior, spring and summer are the times of greatest stream flow in the Fraser basin. When spring or early summer rains coincide with peak periods of snowmelt, flooding can be and has been a significant problem in the basin, especially in the lower Fraser subbasin. Even in nonflood periods, the amount of precipitation and stream flow concentrated in the lower Fraser valley has contributed to drainage problems there. Millions of dollars have been invested in construction and maintenance of dyke works, with financial assistance from the federal and provincial governments, to keep streams in the lower river area within their banks and to preserve agricultural lands and building foundations from seepage.

The interior portion of the basin is drier, and even subject to occasional drought. Nonaboriginal development of water use in the basin began with irrigation in the interior plateau in the mid-1800s. Farming and ranching in the interior, along with extractive industries such as timber and mining, sometimes compete for relatively scarce surface water supplies, particularly towards the late summer and autumn.

The river basin is also rich and diverse in natural resources. Eleven of the 14 biogeoclimatic zones of British Columbia occur in the basin, where an estimated 512 noninsect animal species live. The Fraser River is a great salmon-producing system, with more than half of Canadian catches of sockeye and pink salmon from the river and its tributaries. For the many First Nations in the basin, fishing is important as both an economic and a cultural pursuit. Combining aboriginal and nonaboriginal commercial and recreational activities, fishing yields an annual return in excess of Canadian $300 million (Marshall 1998).

7.1.3
Water Resource Problems

In addition to the flood hazard mentioned in the preceding section, the principal resource management challenges in the Fraser basin may be summarized as follows.

Toxic Discharges

Although toxic discharges have declined due to municipal sewage treatment plant improvements and the adoption of new technologies at pulp and paper plants in the basin, concentrations of toxic materials (for example chlorinated organic compounds such as guiacols from pulp mills) have accumulated in estuarine fish far downstream from discharge points. Toxic materials have also accumulated in the sediments and biota of poorly flushed streams and in areas adjacent to outfalls. Precipitation is contaminated by heavy metals (for example lead and mercury) as well as polynuclear aromatic hydrocarbons and acidic gases, evidently from atmospheric emissions in the Greater Vancouver area. A large number of lower-volume discharges to the Fraser River carry industrial wastes (FREMP 1996; McGreer and Belzer 1999; Shaw and Tuominen 1999).

Agricultural Pollution

Some lakes in the interior areas of the basin – for example Williams, Loon, and Dragon lakes – are showing nutrient impacts from animal wastes. Intensive agriculture in the Fraser valley has contributed more recently to concerns about groundwater contamination from fertilizer and pesticide applications. In the lower Fraser valley, groundwater has been contaminated by manure, fertilizers, and pesticides, particularly the Abbotsford-Sumas aquifer and the Brookswood aquifer.

Threats to Fisheries

Although fisheries remain a major source of livelihood and income, the resource is increasingly under threat. Eight of 15 streams designated by the British Columbia government as sensitive under the Fish Protection Act are in the Fraser basin. Comparing the most recent decade with the historical record, the number of salmon returning to spawn has decreased in half of the basin streams assessed by the Fraser

Basin Council, while increasing in others. Dyking and drainage in the lower basin area have reduced the extent of estuarine wetlands, which are important to salmon and waterfowl populations. On accessible lakes and streams in the upper basin, in-tensive recreational or sport fishing competes with aboriginal food and commercial fishing.

Water Shortages

Although fewer people live in the interior portions of the basin, their per capita water use is more than twice that of the lower valley. Also, the pulp mills found in the interior and upper basin use more water than any other industry in the basin. Thus, even in those portions of the basin where development is less extensive, water use can reach the capacity of water supplies in dry periods.

Water Use Conflicts

Conflicts over water use and wastewater disposal are most intense in the estuary of the Fraser River. The interests of commercial, recreational, and First Nations fishers compete with one another, as well as with use of the river for river transportation and municipal and industrial waste discharges. Riverside access for shipping conflicts with the desires of contemporary urban dwellers, governments, and developers for waterfront homes and restaurants, river walks, and green space. Effluent from three primary wastewater treatment plants in the region pollutes the water relied upon by fish for habitat, fishers for livelihood, and residents or tourists for recreation. In addition, in several portions of the basin there are competing demands on dams and reservoirs to generate electrical power, reduce flood hazards, and maintain stream flows for fish habitat.

Although there are serious water resource management problems in the Fraser basin, there are also favorable situations not found in some of the other cases included in this book. There remain undeveloped headwaters with pristine water quality. The main stem of the Fraser River has never been dammed, and probably will not be in the future due to its designation in Canadian policy as a Heritage River. The large size of the river basin, the large volume of flow on the main stem in normal years, and the fact that urban development has been concentrated mainly near the river mouth have reduced the negative impacts on the basin as a whole.

7.2
Decentralization Process

7.2.1
Pre-reform Arrangements for Water Resource Management

To gain an understanding of the context in which the reform process in the Fraser basin took place, it is instructive to consider intergovernmental relationships as they operate in Canada. The federal system of Canada gives provincial legislatures power over natural resources, including inland waterways and lakes. Thus, most major water uses in the Fraser basin operate under permits or licenses issued by

British Columbia authorities, operating primarily from the provincial capital in Victoria. However, the federal government has power over, or plays a significant role in, interprovincial and international trade, navigation and shipping, conservation and protection of oceans and fisheries, and water on federal lands, in national parks, and in First Nations communities. It is not therefore possible to say simply that natural resources (particularly water) are a provincial matter in Canada. As Table 7.1 indicates, water resource management in British Columbia involves all orders of government – federal, provincial, local, and First Nations – and involves many agencies and organizations.

Table 7.1. Principal agencies and functions in the Fraser basin. *Source:* Calbick et al. (2004)

Order (level) of government	Principal agencies	Primary purposes	Formal authority
Federal (Canada)	Environment Canada	Regulates and monitors water quality	Environmental Protection Act
	Department of Fisheries and Oceans	Manages anadromous fish stocks and habitat	Fisheries Act
	Infrastructure Canada	Matching funds for water/wastewater treatment plants	Numerous
	Department of Canadian Heritage	Administers the Heritage River program	Canadian Heritage Act
Provincial (British Columbia)	Ministry of Water, Land and Air Protection	Regulates water resource quality and quantity	Waste Management Act
	River Forecasting Center (now part of Ministry of Water, Land and Air Protection)	Forecasts water quantity conditions	
	Ministry of Sustainable Resource Management	Develops water resources, provides guidance and leadership on water policy and planning	Numerous
	Land and Water B.C., Inc. (until September 2005)[a]	Manages surface water allocation by issuing licenses	Water Act
	Ministry of Health Services	Regulates drinking water quality	Drinking Water Protection Act
	B.C. Hydro	Hydropower generation	Hydro and Power Authority Act
Local (numerous)	Municipalities, regional districts, improvement districts	Water treatment and distribution; wastewater collection, treatment, disposal services	Local Government Act
First Nations (numerous)	Indian and Northern Affairs Canada	Administers potable water supplies on reserves	Indian Act

[a] In June 2005 the British Columbia government announced that effective September 2005 the activities of this agency would be redistributed among other ministries and departments. The reorganization is still under way at the time of writing.

The matter is further complicated by the horizontal distribution of water-related responsibilities within each government. This has not, however, prevented governments and their personnel from working together on important initiatives and sustaining them over time. One example is the Canada-British Columbia Water Quality Monitoring Agreement, established in 1985 and still in effect. Under this agreement, government personnel have performed biweekly sampling and reporting of results on the presence of ions, nutrients, trace metals, and an indicator of industrial discharges from a number of locations in the basin.

It was this balance of federal and provincial responsibilities, with considerable opportunities for intergovernmental cooperation in areas of overlap, that formed the background to the reform process during the 1990s. Such intergovernmental cooperation in natural resource management activities can be fruitful but entails high coordination costs, and does not necessarily bring other basin stakeholders into a process of information development and sharing, communication, and decisionmaking. The Fraser River Estuary Management Program (FREMP), initiated in 1985, demonstrates the benefits and the drawbacks of the interagency partnership approach. This program will be considered in the next section as it was a major stimulus to reform throughout the rest of the Fraser basin.

7.2.2
Impetus for Reform

FREMP emerged from a study initiated in 1977, went through some organizational modifications in the mid-1980s, and still operates today (Calbick et al. 2004; Dorcey 1990). As it approaches the Pacific the Fraser River diverges into three major distributaries, which flow through the Greater Vancouver metropolis before emptying into the Strait of Georgia. Port facilities, river traffic, salmon and sturgeon fisheries, urban wastewater and stormwater disposal, and a host of other interests and uses converge in this estuarine region. Concerns about the degradation of water quality in the estuary, protection of fish habitat and the livelihoods of fishers, and preservation or even expansion of river transportation as an essential element of the regional economy rose in the 1970s, along with governmental policy interest in coastal zone management and comprehensive basin planning.

Thus began the Fraser River Estuary Study in 1977 under a federal-provincial agreement, guided by a federal-provincial steering committee. After an initial three-year phase, the scope of the study and the composition of the steering committee were broadened. The steering committee became the Fraser River Estuary Planning Committee, which later published a report outlining a number of options and actions to be taken for improving conditions and accommodating the multiple and sometimes conflicting uses within the estuary. This report was revised and adopted by the federal and provincial governments and two port authorities in 1985 as the Fraser River Estuary Management Program (FREMP). The program's executive structure involves considerable representation from all levels of government. Through FREMP, coordinated review of permit applications within the estuary has been improved among governments, and detailed management plans have been developed and agreed for major environmental subsystems within the estuary.

FREMP has been a success, but it involved intensive and sustained collaboration among multiple federal, provincial, and local governments. It has been criticized, furthermore, for lacking a formal place for nongovernmental organizations. In 1990–1991, therefore, when efforts began to focus on developing the Fraser River Action Plan for improving the conditions of the entire Fraser River, attention was given to the idea of a basin management board responsible for planning and executing projects, with input and funding from governmental agencies and with participation by First Nations and nongovernmental bodies. From this concept emerged the Fraser Basin Management Program.

7.2.3
Reform Process

The five-year Fraser Basin Management Program (FBMP), run by the Fraser Basin Management Board (FBMB), was inaugurated in 1992. The FBMB was intended as a multiorganizational, multi-interest committee with the purpose of encouraging consensus-based decisionmaking about basin activities and with a commitment to employing consensus decisionmaking itself. Table 7.2 shows the context of the FBMP within the overall chronology of institutional development within the basin.

The FBMB was the first basin-scale organization in the Fraser basin. The board was composed of 19 members: 12 from the four orders of government in the Canadian political system (federal, provincial, local or regional, and First Nations), six from

Table 7.2. Chronology of institutional development in the Fraser basin

Year	Programs and organizations	Related events
1977	Fraser River Estuary Study, guided by federal-provincial steering committee (1977–1980) and Fraser River Estuary Planning Committee (1980–1984)	
1985	Fraser River Estuary Management Program (FREMP)	
1990	Fraser Cities Coalition formed, which establishes Fraser Basin Start-Up Committee	Vancouver and Prince George mayors challenge each other to clean up the Fraser
1992	Fraser River Action Plan	Federal Green Plan provides funding for Fraser River Action Plan
1992	Fraser Basin Management Program (FBMP) under Fraser Basin Management Board (FBMB)	
1993		FBMB adopts strategic plan
1995		First state of the basin report
1997	Fraser Basin Council	Charter for Sustainability approved
2000		First state of the basin conference
2003		First snapshot on sustainability indicators report

nongovernmental organizations representing economic, environmental, and social interests in the basin, and one appointed neutral chair. In the early stages of its existence the board put considerable effort into the development of communication and organizational skills, given the varied backgrounds and levels of expertise of its members, their differing views on the board's mission and scope of authority, and the fact that there was no basin-scale predecessor organization from which it could evolve and adapt (Calbick et al. 2004:59–60).

In 1993 the FBMB adopted a strategic plan for the FBMP, centered upon a set of wide-ranging principles related to such matters as conservation and prudent management of resources, equal and fair access to information and decisionmaking processes, incorporation of aboriginal interests and concerns, and encouragement of integrated and innovative approaches to basin planning.

The FBMB engaged professional staff and leased its own office space, rather than relying on the participating governmental agencies to provide these services. This has contributed to the confidence of participants in the transparency of information generation and sharing, and fostered perceptions of independence and legitimacy for the FBMP that reinforced the commitments of nongovernmental and First Nations representatives. Of equal significance was the decision in 1995 to hire regional coordinators and place them in the four main regions of the basin. The move, which has been continued by the Fraser Basin Council (which succeeded the FBMB in 1997 – see below), greatly improved information flow, and encouraged local stakeholders to renew their commitment to basin planning and management efforts.

Another important practice initiated by the FBMB and continued under the Fraser Basin Council has been the development of a set of sustainability indicators and the regular publication, since 1995, of a state of the basin report, accompanied by a briefer "report card", grading progress in the basin on some of the more critical issues (Calbick et al. 2004:66). A major report, Snapshot on Sustainability: State of the Fraser Basin, was published in 2003. These reports are indicative of the explicit incorporation of assessment methods and progress reporting into the basin governance and management structure, providing data that can be monitored over time to document changes in basin conditions.

In 1997 the FBMB was replaced by the Fraser Basin Council, with the Fraser Basin Society acting as legal custodian of the council's constitution and bylaws. The Fraser basin is thus unique among the case studies in this book in having a pair of nongovernmental organizations as its principal management and governance institutions. The Fraser Basin Council is a planning and management body composed of 36 representatives drawn from diverse geographical and sectoral communities within the basin, as well as from the four orders of government.

In the same year a landmark document appeared, the basin Charter for Sustainability, published by the Fraser Basin Council but initially prepared by the FBMB as the five-year FBMP neared its completion (Fraser Basin Council 1997). The most striking aspect of the charter is the tremendous breadth that is given to the concept of basin sustainability. The document's vision statement embraces four directions, namely understanding sustainability; caring for ecosystems; strengthening communities; and improved decisionmaking. These directions are guided and informed by 12 principles, as follows:

- Mutual dependence
- Accountability
- Equity
- Integration
- Adaptive approaches
- Coordinated and cooperative efforts
- Open and informed decisionmaking
- Exercising caution
- Managing uncertainty
- Recognition
- Aboriginal rights and title
- Transition takes time

The implications of this broad agenda will be discussed further below.

7.2.4
Current Situation

In the short period of its existence the Fraser Basin Council has become a highly effective organization with a visionary and wide-ranging agenda. Among the principal goals of the Fraser Basin Council has been the promotion of a perspective of interdependency and relationship among residents and communities throughout this very large basin. (For example, upstream pulp mills contaminate fish hundreds of kilometers downstream in the estuary; air pollution from the Greater Vancouver area blows eastwards and contaminates the inland reaches of the Fraser valley.) Such an approach is consistent with the council's broad view of sustainability as embracing a wide range of social, economic, and environmental elements.

This breadth of concept is also reflected in the Fraser Basin Council's programs and finances. Its basinwide programs include flood hazard management, strengthening communities, invasive plant strategy, sustainable fisheries strategy, and a First Nations action plan. Regional programs include mine reclamation, the Greater Vancouver sustainable region initiative, agricultural nutrient management, lake water quality, and river environment enhancement (Calbick et al. 2004:71–81).

This diverse agenda, for such a young organization, reflects a major shift in financing away from reliance on funding support from government agencies and towards project funding. Such funding comes from public and private organizations that contract with the Fraser Basin Council to carry out various activities – perform a study, organize an event, administer a program. Between 1998 and 2003 the budget proportion derived from project funding grew from 4 percent to 36 percent, while that from government agencies declined from 95 percent to 51 percent (though it is important to note that the actual amount from this source has not declined; the council's revenue has roughly doubled in the same period, with project revenue accounting for most of the increase).

7.3
Application of Analytical Framework

Application of the analytical framework (Chap. 1) to the Fraser basin case yields the following observations.

7.3.1
Contextual Factors and Initial Conditions

At least three factors about the Fraser basin setting contribute to its prospects for successful basin management. One is the level of economic development of the nation, and a second is the level of economic development of the basin. Canada generally, and the Fraser basin particularly, are prosperous enough that policymakers and stakeholders have some resources to devote to research, institution building, meetings, projects of environmental improvement, and monitoring and assessment. A third is the initial distribution of resources among basin stakeholders: the vast majority of land and water resources in the Fraser basin are held as a public trust by the province of British Columbia or the government of Canada and are used by private individuals under lease arrangements. This situation has allowed institutional arrangements to develop in the basin under conditions where no single interest or sector of basin users enjoys across-the-board priority or privilege in its claims to resource use – in other words, urban uses are not all privileged over rural uses or vice versa, mining over agriculture or vice versa, and so on.

One aspect of the initial conditions in the Fraser basin presents a challenge: the presence of social and cultural distinctions among basin stakeholders. The claims and title of aboriginal peoples (First Nations) versus the established economic and political power of the nonaboriginal descendants of European settlers has been a difficult issue of long standing throughout Canada, and this is certainly true of the Fraser basin. First Nations and nonaboriginal residents have had difficulty working together, understanding one another, and forging institutional arrangements for joint problem solving. The First Nations issue is never far from the surface of any natural resource issue in the Fraser basin or elsewhere in Canada.

7.3.2
Characteristics of Decentralization Process

In the case of the Fraser basin, as in some of the other cases described in this book, it has not always been clear that there has been a decentralization process in the strict sense of the term. The construction of basin-scale institutional arrangements in the basin appears to be as much or more a matter of integrating already decentralized organizations and jurisdictions rather than decentralizing previously centralized ones. Nevertheless, some of the considerations in this category are definitely relevant to the Fraser basin case. One is the extent of central government recognition of local-level basin governance, which has been extraordinarily positive. The Canadian national and British Columbian provincial governments joined in the predecessor organizations in the basin (the Estuary Steering Committee, the FBMB); they funded the Fraser River Action Plan from 1992 through

1998; and they have been original and consistent members of the Fraser Basin Council and have supported it financially.

Another factor has been the consistency of that support through changes of government and administration at both the provincial and federal levels. Although Fraser Basin Council members and staff are always alert to the possibility that electoral changes of government might bring shifts in commitment, thus far the institutional arrangements for Fraser basin management have maintained support from both levels through electoral changes. It remains to be seen how the council will cope with recent changes in its own leadership structure. Losing strong and committed champions who possess well-developed managerial skills as well as political acumen can sometimes disrupt agency activities.

7.3.3
Central-Local Relationships and Capacities

There are a number of favorable factors operating in this category. The financial resources and the financial autonomy of the Fraser Basin Council are quite strong, though they remain an important concern of the members and staff. Through the Fraser Basin Society and the council's own bylaws, the council members have demonstrated the ability to create and modify the institutional arrangements with which they work, as exemplified by the adoption of the Charter for Sustainability, and the transformation of the Fraser Basin Management Board into the Fraser Basin Society and Fraser Basin Council structure that exists today. As suggested above, however, the water rights system is something of a mixed bag – on the one hand, the arrangements governing rights to water and land use allow for considerable management flexibility, but on the other, the control of groundwater resources is particularly weak and represents a current and future vulnerability in the water resource management aspects of the overall basin sustainability effort.

7.3.4
Basin-Level Institutional Arrangements

The strongest features within this category are the availability of a basin-level governance body (the Fraser Basin Council), the recognition of subbasin communities of interest through the composition of the council (by including regional representatives and through its employment of regional coordinators), and the institutionalization of regular monitoring of basin conditions by means that are trusted by resource users. The council was designed quite deliberately to provide information sharing and communication among basin stakeholders, to provide means for basin stakeholders to enter into agreements to take actions for improvement of basin conditions, and to resolve conflicts. The one variable in this category that is not entirely favorable is related to the clarity of institutional boundaries – while the Fraser Basin Council has emerged as the paramount deliberative body in the basin, in its capacity as a nongovernmental organization funded through a nonprofit society, the council generally cannot turn its decisions and plans into actions. It usually hands off projects to other (usually governmental) entities for implementation, and at times even the council members are not entirely clear what actions are within the council's scope.

7.4
Performance Assessment

7.4.1
Stakeholder Involvement

A large number of different stakeholder groups are affected by, or are trying to address, the resource management issues in the Fraser basin. These include federal, provincial, and local government organizations; First Nations communities; port authorities; natural resource-based industries, including mining, forestry, and pulp and paper manufacturing; other industry; agriculture; commercial fishing; recreation and tourism; and environmental organizations, including those affiliated to the British Columbia Environmental Network. All of these stakeholders and their activities are affected by, and affect, water quality and other environmental conditions within the basin.

Each of these groups has its own views on the nature, functions, and benefits to them of the Fraser Basin Council. Federal agencies have been willing to fund, send representatives to and cooperate in programs with the council because it allows them to influence resource management issues that might otherwise be beyond their constitutional authority, and because the council is an organization to which the federal agencies can hand off problems and concerns for investigation and discussion. The council also allows the agencies to satisfy statutory and regulatory obligations for public participation in basin management programs.

Provincial ministries and their representatives find in the Fraser Basin Council a means to break out of the substantial interagency fragmentation of water resource responsibilities at the provincial level, overcome budgetary limitations on their resources, and engage greater participation. As with the federal agencies and their representatives, these benefits of council participation suffice for provincial ministries to maintain their annual financial support of the council.

Furthermore, the council's consensus approach to decisionmaking has helped to assure and maintain federal and provincial agency representation. By definition, agency representatives serving on a consensus-based group are shielded from being in the position of belonging to an organization that takes positions contrary to federal or provincial policy. The council would be unable to come to consensus on any such position.

Other Fraser Basin Council participants (local government, First Nations, and regional and sectoral representatives) get access to good information, a chance to raise issues and concerns in a forum where federal and provincial representatives are listening, and opportunities for coalition building to enhance their political influence. Here too the consensus approach provides an incentive to participation, since it effectively places these stakeholder representatives on an equal plane with representatives from federal and provincial agencies that have the constitutional and statutory authority as well as the budgetary resources most local, aboriginal, and sectoral representatives lack. The Fraser Basin Council itself has no constitutional or statutory authority to execute decisions on resource management policy, but its structure and operation place the officials who have that authority at a table (literally and figuratively) with other stakeholders in the basin and this is an important element of their continued commitment to it. Not to be overlooked is the spirit of cooperation generated by the social closeness, and genuine friendship, that has grown out of the interaction between the individuals who serve on the council.

7.4.2
Developing Institutions for Integrated Water Resource Management

The breadth of the Fraser Basin Council's agenda (see Sect. 7.4.3) is reflected in, and reflects, some of its organizational characteristics. As noted earlier, the council is composed of 36 members, an expansion from the predecessor FBMB. Furthermore, Fraser Basin Council seats are deliberately distributed so that no sector of basin interests or level of government has a majority of members and so nonwater as well as water-related basin interests are represented. In practice, more than 36 interests are represented at council meetings, as several members wear a number of hats – a regional representative may be a rancher, or work in forestry, for example.

A striking feature of the make-up of the council is the presence of eight First Nations representatives, reflecting the sustained effort to incorporate aboriginal along with nonaboriginal interests in basin water resource management.

The Fraser Basin Council has continued the FBMB practice of maintaining consensus-based decisionmaking. Each policy recommendation or programmatic involvement of the council has to be acceptable to all members; otherwise, the matter is continued for further discussion and refinement, or dropped.

As noted earlier, the Fraser Basin Council maintained the earlier FBMB practice of employing regional coordinators in addition to the staff in Vancouver. The council has divided the Fraser basin into five regions, shown in Fig. 7.2, which are the basis not only for the assignment of regional coordinators but for the designation of the 10 regional representatives on the council.

7.4.3
Effectiveness and Sustainability

As previously explained, an outstanding feature of the approach of the Fraser Basin Council is its broad view of the concept of sustainability, as outlined in its Charter for Sustainability, and the way in which this outlook is woven into its activities and the way they are financed. The shift in emphasis from government funding to project funding has significantly widened the council's financial base and has enabled it to become involved in a great variety of projects, some only indirectly related to water resource management.

In addition to these projects, the Fraser Basin Council has continued and expanded the public information and outreach programs started by the FBMB in the mid-1990s. The council publishes annual reports, state of the basin reports and snapshots on sustainability, and holds biennial conferences, all focused on basin conditions (Calbick et al. 2004:82–106).

Positive or negative changes in basin conditions may be assessed by reference to a selection of natural resource-related measures or indicators, namely fisheries;

Fig. 7.2. Fraser Basin Council designated regions of the Fraser basin. *Source:* Fraser Basin Council (*www.fraserbasin.bc.ca/ regions/index.html*)

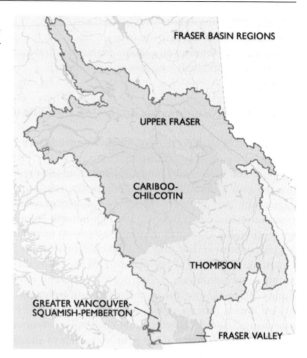

pollution and toxic discharges; and water quality monitoring (though not all changes or improvements result from Fraser Basin Council programs or activities).

Fisheries are recovering from a period of decline: comparing the most recent ecade with the historical record, the number of salmon returning to spawn has increased in half of the basin streams assessed.

As regards pollution and toxic discharges, measured concentrations of most pollutants in the main stem of the Fraser River have not exceeded water quality guidelines. The exceptions are iron and copper (which may be exceeded because of naturally high background levels in the basin) and the industrial wastewater indicator adsorbable organohalides downstream of pulp and paper mills. In the basin as a whole toxic discharges have declined due to municipal sewage treatment plant upgrades and the adoption of new technologies at pulp and paper plants. Lower Fraser River bottom sediments have shown improving trends in lead concentrations. Restrictions on fish consumption from the Fraser River upstream of Hope were lifted by the British Columbia government in 1994, and on the Thompson River in 1995.

A number of studies and water monitoring projects, several undertaken by the Department of Fisheries and Oceans in partnership with Environment Canada, have been carried out since the inception of the Fraser River Action Plan in 1992, and have given a much clearer picture of the extent, sources, and potential mitigation of pollution in the Fraser River. These studies have yielded important baseline information concerning contaminant levels in the river, and have led to a greater understanding of how the river ecosystem functions, particularly with regard to factors that affect its salmon (Calbick et al. 2004:57).

The sustainability of the work of the Fraser Basin Council is enhanced by the stability and vocational aspirations of its staff, many of whom express personal commitment to the sustainability principles that the council articulates and espouses, and see employment with the council as an opportunity to put these principles into practice. That motivation is reinforced by the belief that the Fraser Basin Council is an organization that is truly making a positive difference, in ways that go beyond the opportunities that might be available with a more traditional nongovernmental organization. On the other hand this level of commitment may tempt staff and council members to pursue projects that seem only tangentially related to basin management concerns, and it is acknowledged that there have been occasions when funding opportunities associated with involvement in one or another project have stretched the council's own broadly defined scope and agenda rather far.

In 2002 the Fraser Basin Council employed a consulting firm to interview council members, staff, and external observers to assess the council's own performance and effectiveness (SALASAN Associates Inc. 2002). The report, which echoes a number of the points raised above, found that involvement in the council's work had been a satisfying experience for participants, and that the council's status as a nongovernmental body and its broad representation are extremely helpful in addressing issues that cross agency domains and jurisdictional boundaries and promoting inclusion of a wide range of perspectives. On the other hand, the council occasionally becomes involved in issues that are not clearly related to its charter and runs some risk of loss of focus.

7.5
Summary and Conclusions

7.5.1
Review of Basin Management Arrangements

Since one of the most distinctive features of this case is the role of the Fraser Basin Council as a nongovernmental organization, some concluding comments are in order about how well that has worked in the Fraser basin and its possible implications elsewhere. It can certainly be argued that the nongovernmental model reduces some of the bureaucratic turf battles that one would expect to be associated with placing basin management responsibility in an existing agency, or creating an agency that would have authority and responsibilities transferred from or overlapped with existing agencies. The nongovernmental approach also fits well with a federal system such as Canada's, since it provides a means of crossing jurisdictional boundaries among levels of government in a context where a constitution divides authority and one level of government is not entirely superior or subordinate to the other. It is also suited to a common-law cultural context where private organizations are free to do anything that is not forbidden by law, and to take actions (including the raising and distribution of funds) up to the limits of public authority.

Furthermore, the nongovernmental approach in the Fraser basin has allowed for the integration of First Nations and private stakeholders in ways that more traditional intergovernmental programs have often found difficult if not impossible. It has served as a good forum for information generation and sharing, since there is less concern over who owns the information. A nongovernmental organization has the boundary flexibility to cover the whole basin (which no local government can do) but not more than the basin (as would be the case for a provincial or federal agency). As already noted, a nongovernmental organization of the Fraser Basin Council type also provides good political cover for agencies, who can justify actions that might otherwise be unpopular with some constituency.

The approach also has its weaknesses and drawbacks. Most important is the fact that the Fraser Basin Council is generally unable to implement the plans and programs it agrees upon, and must hand them off to others – usually governmental agencies – for actual performance. This limitation means that matters on which the council has made recommendations do not always get done or get done swiftly or without modification by the implementing agencies. A more vigorous advocacy role, prodding governments or other bodies for action, has its own risks, however, as one of the council's most important assets is its reputation for neutrality.

Other vulnerabilities include the fact that the Fraser Basin Council's consensus decisionmaking approach, though helpful in a number of respects, is also time consuming and can be frustrating. As a nongovernmental organization financially reliant upon goodwill contributions and funded projects, the Fraser Basin Council is subject to enough budgetary uncertainty (despite the consistency of governmental contributions to date) to limit its ability to commit to long-range projects. An organization in such a position is also continually vulnerable to "mission creep", the temptation to follow the money that is available for projects that may be beyond its primary concerns and interests.

On balance, the approach represented by the Fraser Basin Council has worked well in this setting as a means of bridging fragmented public authorities and integrating indigenous and other private stakeholders. It has succeeded so far in preserving a reputation for objectivity and avoiding widespread perception of bias, and in building a more diverse financial base. The council's structure, agenda, and performance are key reasons the Fraser basin has proved to be a valuable addition to the research cases reviewed in this book.

7.5.2
Future Prospects

As has been shown, the water resource management system applied in the Fraser basin, with the Fraser Basin Council as the lead agency, has proven to be an effective means of coordinating action between an extremely varied range of stakeholders over a large diverse area. It has carried forward the positive attributes of its predecessor the FBMB, and it is likely that its future effectiveness will be sustained by building on its current assets – its key intergovernmental role, its ambitious and integrated agenda, its accommodation of disparate stakeholder factions, its committed staff, its recognition of the value of clear information flow and communication, and its emphasis on sustainability. In this last respect, the increasing use of project funding has given added financial stability to the council's operations. It remains to be seen, however, how well the Fraser Basin Council will balance the different elements of its diverse agenda as its budget and its role in basin water resource management expand.

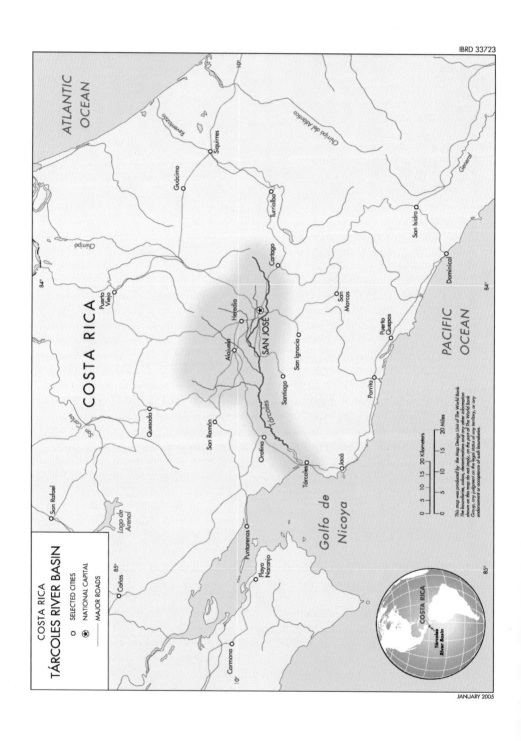

IBRD 33723

ATLANTIC OCEAN

Reventazón

Chirripó del Atlántico

General

Squirres

Guácimo

Turrialba

Cartago

San Isidro

Chirripó

84°

COSTA RICA

Puerto Viejo

Heredia

Alajuela

SAN JOSÉ

San Marcos

Dominical

84°

San Carlos

Quesada

San Ramón

Santiago

San Ignacio

Puerto Quepos

PACIFIC OCEAN

Orotina

Tárcoles

Parrita

Lago de Arenal

San Rafael

Tárcoles

Jacó

Golfo de Nicoya

85°

Cañas

Puntarenas

Playa Naranjo

85°

Carmona

10°

COSTA RICA
TÁRCOLES RIVER BASIN

o SELECTED CITIES

⊛ NATIONAL CAPITAL

— MAJOR ROADS

0 5 10 15 20 Kilometers

0 5 10 15 20 Miles

This map was produced by the Map Design Unit of The World Bank.
The boundaries, colors, denominations and any other information
shown on this map do not imply, on the part of The World Bank
Group, any judgment on the legal status of any territory, or any
endorsement or acceptance of such boundaries.

COSTA RICA

Tárcoles River Basin

JANUARY 2005

Chapter 8

Costa Rica: Tárcoles Basin

W. Blomquist · M. Ballestero · A. Bhat · K. E. Kemper

8.1
Background

8.1.1
Introduction

With its mountainous spine and numerous valleys, Costa Rica contains 34 river basins. The Tárcoles basin – the drainage area of the Río Grande de Tárcoles – is located in the west-central portion of Costa Rica and extends from the mountain ranges in the middle of the country to the Pacific coast (Fig. 8.1). The basin is of great economic importance to Costa Rica, with much of the urban, industrial, commercial, and agricultural activity of the country concentrated within its borders. This has placed immense pressure on the water resources of the basin, with water quality issues proving particularly problematic.

Compared to some of the other case studies in this book, integrated water resource management in the Tárcoles basin has emerged relatively recently with the establishment of a basin commission, the Commission for the Río Grande de Tárcoles Basin (Comisión de la Cuenca del Río Grande de Tárcoles; CRGT), in the early 1990s. This case has been extremely valuable because the formation of the basin commission was a locally initiated action that occurred in the fairly recent memory of many individuals who are still actively involved in water and government, providing an opportunity to explore the early life cycle of a river basin organization and some of the factors linked to its origin, early growth, and recent stagnation.

8.1.2
Basin Characteristics

The Río Grande de Tárcoles, a river of 111 kilometers' length emptying into the Pacific, is formed by the confluence of the Alajuela, Grande, and Virilla Rivers toward the middle of the basin. Costa Rica as a whole has abundant precipitation, ranging from 1 200 to 7 000 millimeters per year, producing plentiful runoff. The river basins on the Pacific side, such as the Tárcoles, tend to be somewhat drier, with a noticeable reduction of river flow during the dry season of the year. Still, precipitation in the Tárcoles basin ranges from 948 to 5 409 millimeters per year, with an annual average of 2 364 millimeters. Flooding is a recurring problem in the basin, as in most of the river basins of Costa Rica.

There are three distinct subareas within the Tárcoles basin – an upper area that corresponds with the watershed of the Virilla River (about 40 percent of the total

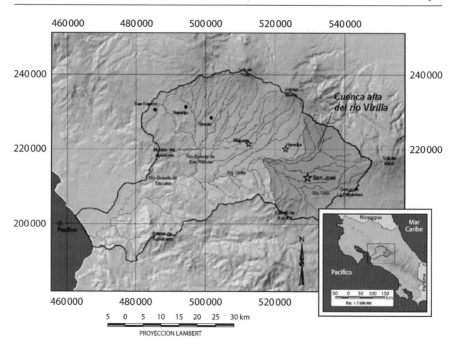

Fig. 8.1. Location of the Tárcoles basin within Costa Rica. *Source:* Ballestero 2003

basin area), a middle area that corresponds with the watershed of the Grande River (34 percent), and a lower area along the course of the Río Grande de Tárcoles below the confluence of the Virilla and Grande Rivers (26 percent). The middle basin is semirural with some population centers, and the lower basin is mostly rural. The upper basin contains about 80 percent of the population of the basin, with large concentrations of both urban population and industry. It also contains the largest aquifer systems in the basin – the Barva and Colima aquifers, which are layered aquifer systems that supply most of the groundwater used by industry and the urban population.

Relative to the rest of Costa Rica (and in light of the large number of river basins there), the Tárcoles basin is fairly large at 2155 square kilometers total area, and the Virilla and Grande watersheds are two of the largest in the country. But these geographic dimensions do not capture the great significance of the Tárcoles basin to Costa Rica. With only 4.2 percent of the land area of the country, the basin is home to half of its population (approximately 2 million out of 4 million), contains 80 percent of its industry, 80 percent of the vehicles, and more than half of the registered wells.

The cities in the upper basin have grown into a large metropolis at the center of Costa Rica known as the Gran Área Metropolitana (Greater Metropolitan Area), which includes San José, the nation's capital, and three other cities. The Greater Metropolitan Area is also the transportation center of the country, with the national highways from other regions of Costa Rica converging and intersecting in the Tárcoles basin. The area is also crucial to Costa Rica's important and growing tourism industry, as most tourists pass through the international airport in Alajuela and stay in the area for at least a portion of their time in the country.

Despite the growth of urban and industrial centers within the basin, 37 percent of the land use remains in crops and pasture. Coffee farming, other crops with and without irrigation, dairy farming, and livestock ranching occur throughout the basin.

8.1.3
Water Resource Problems

Although precipitation is abundant, the concentration of people, industry, and agriculture in the Tárcoles basin translates into significant and growing water resource problems. Those problems are exacerbated by institutional arrangements governing water management at the national level, and by financial constraints that have kept infrastructure development within the basin from keeping pace with economic development and population growth. The 2001 State of the Nation report identified the vulnerability of water resources and water quality as Costa Rica's biggest environmental concern, with the Tárcoles basin as the principal focus for that concern.

Sewage, Solid Waste, and Water Pollution

It is a striking fact that in the Tárcoles basin, with its 2 million inhabitants and abundant economic activity, 96 percent of domestic and industrial waste is untreated, much ending up in the basin's streams and rivers. Industrial wastes contribute contaminants in addition to the sewage that emanates from households in the basin. Agribusinesses, especially food and coffee processing industries, are prevalent throughout the upper and middle basin areas, and the upper basin contains 29 chemical and alcohol manufacturing facilities. Although national law requires treatment of industrial wastes, most industries still do not have facilities in place. The surface waters of the basin are also a depository for rubbish, which, in addition to the obvious negative aesthetic impact, aggravates flooding problems where it chokes off stream channels, and adds a significant nuisance and expense to the operation of hydroelectric power facilities.

Groundwater in the basin is also vulnerable to quality degradation. Septic tanks serve 68.5 percent of households and businesses in the basin, so those that are not discharging to surface waters are discharging to the ground. Application of fertilizers, especially in high concentrations for intensive coffee farming, is associated with nitrate contamination of groundwater in the upper and middle basin areas (Reynolds-Vargas and Richter 1995).

The polluted surface water already requires significant treatment prior to use, though the fact that many of the smaller water treatment facilities lack the financial and technical capacity to provide acceptably clean water raises public health concerns. It is estimated that 31 percent of the population in the basin receives untreated water.

Deforestation

Deforestation of the upper and middle basin areas continues despite serious national efforts to arrest it. From 1992 to 2000, the forested portion of the basin was reduced from approximately 66 000 hectares to 38 000 hectares. The same period saw a 15 000-hectare reduction in farmland due to conversion of land from agricultural to urban uses. These processes have accelerated soil erosion, reducing the capacity of hillside and valley soils to absorb and retain water, thus aggravating both flooding problems and dry season water scarcity.

Increasing Demand for Water

Urban and industrial water demands have grown on top of the already significant agricultural water use (an estimated 60 percent of consumptive use is for irrigated agriculture). This has led to shortages or uncertain availability of high-quality water supplies for newer, high-value sectors of the economy such as tourism, recreation, and fishing. Some water is even imported to the upper basin area from the neighboring Reventazón basin for use in the Greater Metropolitan Area.

Institutional Context

The institutional capacity to deal with these problems is sorely lacking. A national water policy was ratified by the Government Council in January 2006. In March 2006, the preparation of the first national water plan started. Both form part of the National Water Resources Management Strategy, financed through the Inter-American Development Bank (IDB). Accordingly, these institutional changes are very recent. The existing national Water Law dates from 1942, and modified the first regulation of 1884; a proposed new Water Law has been under consideration in the Costa Rican legislature. The new National Water Resources Management Strategy will hopefully address the weaknesses that exist in the current institutional setup. For instance, the system of water concessions in the country has significant gaps that contribute to uncertainty about water availability, since it is hard to tell what uses are occurring already in the basin and to what extent total water demand exceeds or is exceeded by available supply. For example, the largest hydroelectric power producers are not included in the concession system.

   Tariffs on agricultural water use are based on land area rather than on the volume of water used, providing little incentive for farmers to replace or upgrade aging and inefficient gravity-fed systems. The entire system of water charges fails to provide enough revenue to maintain infrastructure within the basin, let alone support needed improvements such as water treatment plants.

8.2
Decentralization Process

8.2.1
Pre-reform Arrangements for Water Resource Management

Costa Rica has historically had a highly centralized form of government. Decentralization efforts with regard to a number of public services have been isolated, and sometimes only temporary. River basin management has followed the same pattern: there has been no overall, nationwide effort to decentralize water resource management to the basin level in Costa Rica. The creation of the CRGT was the first effort in Costa Rica to establish a structure for basin management. Interestingly, it represented more of a bottom-up than a top-down approach to establishing a basin organization.[1] Since then, basin organizations have been created in a few other basins in the country.

8.2.2
Impetus for Reform

The train of events that led to the setting up of the CRGT in 1994 had its origin in 1991, when the municipality of the central canton of San José, the most influential municipality in Costa Rica, began to pay attention to serious environmental problems that were apparent in the capital city, especially water pollution, the dumping of solid and liquid waste in the rivers, and the recurring problem of rivers and streams overflowing their banks, which had often caused serious tragedies. It was unusual for municipalities to take action on such matters, so while San José's efforts did not violate any formal restrictions on its authority, they did run contrary to stereotypical views of municipal responsibilities.

To pursue its interest in these environmental quality-of-life issues, the municipality of San José initiated a series of activities in 1991 and 1992, of which the following stand out: the Project for the Recuperation of the Río Torres, which focused on controlling the disposal of solid and liquid wastes in the Río Torres, a major upper basin river and one of the most polluted rivers in the Tárcoles basin; a municipal policy directive giving industries located in the canton 24 months to begin treating their liquid waste, followed by a pilot plan, launched by the municipality and the Chamber of Industry, involving a group of businesses that were major polluters; and an urban control plan to regulate land use and urban growth, which included a series of environmental provisions to protect surface water, aquifers, and recharge areas.

By this time, though, officials and staff of the municipality had come to recognize that these environmental matters could not be resolved by actions in only one canton. The situation would have to be approached with a broader view, one that included the entire Tárcoles basin. Visits were made to each of the other 35 municipalities in the basin area to promote their participation and to set priorities. In August 1992, the municipality of San José organized a seminar entitled "The Río Grande de Tárcoles River Basin: Looking Toward the Future", in which a large audience participated in discussions and defined some basic guidelines for coordination to confront the immense task of recuperating the basin.

---

[1] Although the expression "bottom-up" is often used to refer to actions initiated by grass-roots or other civil society entities, it is used here somewhat more broadly, as the CRGT was initiated by local government officials and civic organizations together. This use of the term "bottom-up" still contrasts accurately with a "top-down" decentralization initiative of the central government; CRGT was the former, not the latter.

8.2.3
Reform Process

Following the August 1992 seminar (Sect. 8.2.2), with continued leadership and support from the municipality of San José, the Commission for the Río Grande de Tárcoles Basin, the CRGT, was established, comprising representatives from municipalities, relevant government ministries, state institutes, a university, and private sector organizations. From its beginning, then, the CRGT had a participatory structure that was both interorganizational and interdisciplinary, attempting to connect the most important actors in water resource matters. The structure was nevertheless imperfect, and some of its deficiencies are discussed later.

The CRGT obtained official recognition from the national government through ministerial decree rather than through national legislation, which expedited the process of recognition but later proved problematic, since ministerial decrees are more easily altered or disregarded by subsequent administrations than are national laws. The decree, issued in April 1993 through the Ministry of Environment and Energy (Ministerio del Ambiente y Energía; MINAE), recognized the CRGT's legitimate existence as a collaborative body for information development and sharing, communication, and planning, but it did not transfer to the CRGT any formal responsibility or capacity for undertaking water management projects. By the end of 1993 the commission had 19 members, of which 5 were nongovernmental organizations and 6 were municipalities.

During 1993 and 1994 the CRGT conducted an awareness-raising campaign about river basin management, involving municipalities, nongovernmental organizations, and private and public sectors in a series of workshops and other events. The municipality of San José continued to provide leadership; the CRGT had space in the offices of the municipality, which provided it with staff support. CRGT members and staff participated in training activities.

By 1994 the CRGT was operational and began conducting activities in the basin, with MINAE providing logistical and economic support and devolving to the CRGT some basin planning and study functions, including contracting with the IDB for a large-scale basin study. The headquarters of the CRGT were transferred to space offered by MINAE. MINAE also expressed support for the creation of similar structures in other basins, dividing the country into five watersheds and creating a favorable atmosphere for the deconcentration of services in them. In 1995 the River Basin Program was created for the purpose of determining guidelines for MINAE regulations regarding basins and to formulate national policies for basins.

During the period 1994–1998 the CRGT was very active and became a management model at national and international levels. It was a founder member of the International Network of River Basin Organizations in 1994, and was incorporated into the Latin American Network of River Basin Organizations in 1997. The information and communication functions of the CRGT continued throughout this phase. CRGT members and staff compiled and systematized information on water quality and pollution sources in the basin, studies on basin characteristics, and institutional and legal analyses. Stakeholders were brought together through a number of workshops and seminars.

The CRGT also implemented four major action programs during this period: the Voluntary Plan program, which invited businesses to present voluntary plans to establish waste treatment systems, and eventually attracted the involvement of over 100 businesses; the Ecological Flag program (1994), by which the CRGT awarded "ecological flags" to organizations that had developed and implemented resource protection and recovery activities in the basin; reforestation programs to promote watershed protection, incorporating awareness-raising activities with schools, community organizations, and environmentalists; and the Program for Integrated Management of Natural Resources, a basin-scale coordinated program of protection and recovery for water and other national resources, for which the IDB gave the CRGT the responsibility to supervise the use of US$1 million to design the program and fund the development of feasibility studies.

At the beginning of 1999, a number of factors converged that began to make the normal operation of the CRGT difficult. The newly elected national government adopted a more cautious, centralized approach as the new Minister of Environment and Energy took a more active role in its governance and operations, including the appointment of a new CRGT president. These developments served to highlight the dynamic role of the original president in mobilizing people and resources to make the CRGT an active body, despite its uncertain status as an institution recognized by national government but lacking formal governmental powers. The fragility of this situation was exposed as the CRGT underwent rapid decline.

8.2.4
Current Situation

After a short period (2000–2001) during which the presidency was assumed by a representative of the Union of Local Governments who had been openly critical of CRGT's changed direction (Dulude 2000), MINAE reasserted control with the appointment of a MINAE official as president. Unfortunately this official combined the role of president with other ministerial responsibilities, with the result that the CRGT functioned at quite a minimal level, a situation that tested the commitment of its member organizations and agencies.

The president's office has undertaken some activities in the name of the CRGT, for example research and actions intended to implement economic instruments to promote pollution control, but division of responsibility between the CRGT and MINAE has not been clear. Moreover, there appears to have been little or no interest on MINAE's part in following up on the Program for Integrated Management of Natural Resources, despite the IDB's investment of US$1 million in the 1997–1998 feasibility studies mentioned in the previous section.

The main institutions playing a role in water resource management in the Tárcoles basin are shown in Table 8.1, with a summary of their functions.

8.3
Application of Analytical Framework

The Tárcoles case provides valuable lessons, as the basin commission was formed relatively recently (in contrast to, for example, the longer established organizations of the

Table 8.1. Water resource management institutions and roles in the Tárcoles basin

Management level	Institution	Current water management attributions
National	Ministry of Environment and Energy (MINAE)	– By law, acts as the lead agency of the national water sector – Defines policies and administers national water resources – Grants concessions and authorizes permits for water use or discharge in rivers of public domain – Collects taxes
	Ministry of Health	– Sets standards, controls and monitors quality of water for human consumption, industrial and agribusiness use – Authorizes drainage or discharge of solid and liquid waste that could pollute superficial, underground, or marine waters
	Public Service Regulatory Agency (ARESEP)	– Regulates water supply system and sewer services in harmony with the interests of users and suppliers – Fixes tariffs after consulting interested parties – Quality criteria and control, customer service
	Institute of Aqueducts and Sewers (AyA)	– Designs, finances, builds, and operates water supply systems built after its creation in 1961 – Regulatory authority for water supply systems
	Institute of Electricity (ICE) Subsidiary: National Power and Light Company (CNFL)	– Created in 1949 to plan and conduct the rational development of energy for the country, especially from water resources – Constructs reservoirs for hydropower generation – Its founding law gives it the functions of basin protection and conservation
	National System of Conservation Areas (SINAC)	– Manages protected areas, designated on the basis of their ecological significance or vulnerability
Basin	Municipalities	– Management, administration and maintenance of municipal water supply and sewer systems – Authorization, control and regulation of activities conducted in their areas of jurisdiction – Billing and collection for water service
	Commission for the Río Grande de Tárcoles Basin (CRGT)	– Official recognition via ministerial decree (1993) – Collaborative body for information development and sharing, communication, and planning – Implements some action programs
	Association for Hydrographic River Basins (ASOCUENCAS)	– Consists of members of the commission; is meant to function as the executive arm of the CRGT, but is actually not operational
	Foundation for Urban Development (FUDEU)	– Receives funds and applies them to water resource management issues on behalf of the commission

Guadalquivir and Murray-Darling basins), and occurred as a result of local efforts rather than national political forces. Application of the analytical framework (Chap. 1) highlights some of the key variables associated with the CRGT's progress thus far.

8.3.1
Contextual Factors and Initial Conditions

The process of development in Costa Rica can generally be characterized as successful, with economic growth over the past half century accompanied by a stable political system. However, financial resources for such reform as basin-level integrated water resource management are still limited compared to the size of the task.

Within Costa Rica the Tárcoles basin is by far the most economically developed in the country, and offers the most promise for successful river basin management. Nor do there appear to be substantial cultural or religious differences across groups of basin stakeholders that would dramatically inhibit prospects for cooperation. However, the relative economic and sociocultural complexity of the basin gives rise to its own set of problems that make the process of institutional reform more difficult. In addition, the overall political reluctance to decentralize to lower levels of decisionmaking is reflected in the stalling of the reform process in the Tárcoles basin. More recent organization efforts within Costa Rica, specifically in the Reventazón and Tempisque basins, appear to be making more headway than in the Tárcoles basin, perhaps because resistance is less strong in these smaller and less important basins.

Two additional contextual factors in Costa Rica have shaped the outcomes in the Tárcoles basin so far. The first factor is the relatively large number of separately identifiable river basins in a relatively small country. The Tárcoles is only one of 34 identified basins in Costa Rica, and there is substantial difference of opinion about the appropriate scale at which to organize integrated water resource management. Furthermore, the Costa Rican government has been trying to determine whether and how to coordinate integrated water resource management with other ecological and natural resource policies that are organized on different territorial bases. Costa Rica has systems of national parks and protected areas, which have recently been organized into a national system of 11 conservation areas, but these do not coincide with river basin boundaries or provincial divisions. The inconsistent and hesitant nature of central government support for basin management in the Tárcoles basin may not end until some clear view emerges among national-level policymakers about how to proceed with river basin management organization and what, in Costa Rica, constitutes the appropriate level for water resource management or other natural resource management.

The second factor is a cultural dimension shaped by Costa Rica's physical circumstances and historical evolution. In a humid setting with abundant precipitation, combined with (until recently) a small population and economy to sustain, Costa Rica developed a mostly unwritten but nevertheless widely understood and shared view of water as essentially free and plentiful. Only recently has it become clear that water quantity and quality can be limiting factors in Costa Rica's future economic development and quality of life. The cultural perception of water abundance still contributes to the lack of a sense of crisis, constraining policymakers' ability and willingness to promote sustainable water policy in Costa Rica through more restrictive water rights laws and higher water tariffs, whether organized at the basin scale or otherwise. In the Tárcoles case in particular, this perception contributes to weak controls on water uses and inadequate revenue with which to address the pollution problems in the basin. A considerable public education effort will be necessary as part of any attempt to promote more responsible use of water as a resource of limited quantity and quality.

8.3.2
Characteristics of Decentralization Process

The origins of the CRGT fit closer to the locally initiated effort and mutually desired devolution categories than a shedding or abandonment of central-government responsibilities. Officials at the municipality of San José understood that their water quality problems were related to other municipalities, making a basin approach most appropriate, and took the initiative to create the CRGT. Local leadership initiated the efforts to address basin problems, and central government officials were supportive and provided some help initially and substantial support during the years of peak CRGT activity. However, the tactical decision by basin stakeholders and central government officials to recognize the CRGT by ministerial decree, rather than by a law enacted by the Costa Rica legislature (Sect. 8.2.3), proved unfortunate when, after a change of government in 1998, the ministry balked at carrying out activities not formally authorized by law, placing the central government's support for the Tárcoles commission in doubt.

There were also a number of weaknesses and incongruities in the decree that created the CRGT and its operating regulations that became a burden to the commission later. For example, the decree, by being an instrument of lower rank than a law, cannot confer management responsibilities to the CRGT, creating a gap between the objectives for which it was created and its capabilities to reach those objectives. This reduces its autonomy and seriously limits its scope as a basin organization. To partially address this deficiency, a functional arrangement has arisen whereby two non-governmental organizations – the Association for Hydrographic River Basins (Asociación pro Cuencas Hidrográficas; ASOCUENCAS), which includes CRGT members, and the Foundation for Urban Development (Fundación para el Desarrollo Urbano; FUDEU) – helped provide support for the CRGT, giving it greater operational capability and enabling it to execute projects. In practice, however, the CRGT constitutes a space for meeting and discussion to coordinate the actions that different institutions and social sectors are conducting in the basin.

Also, the decree does not set a budget for the commission's operations and does not define any other method of funding or of providing resources. This becomes a major obstacle to the CRGT assuming a leading role in river basin management. Nor were the responsibilities and roles of the agencies that compose the CRGT defined, and the commission depended largely on the support and goodwill of public and private officials. The representatives of the public sector did not have decisionmaking power and could not make major commitments. Other organizational arrangements proved unsatisfactory: the 36 municipalities in the basin were always represented by delegates from the same 6 municipalities, whereas some rotation of membership might have encouraged greater commitment from and coordination between all basin municipalities.

These deficiencies in the CRGT design have been noticed by Costa Rican officials, but have not led to a systematic reform of the CRGT. The lessons learned have been applied elsewhere instead; for example, in 2000 a law was issued creating Costa Rica's first legally recognized river basin organization: the Commission for the Regulation and Management of the Río Reventazón River Basin.

River basin management in Costa Rica is now widely perceived to be an issue under MINAE's authority and direction. This perception, in addition to the current government's lack of commitment to the issue, has served to marginalize the past efforts of the CRGT. Remarkably, MINAE created a National River Basin Network in 2000 to coordinate national basin policy and promote improved management, data exchange, and awareness raising, and the CRGT – Costa Rica's original river basin organization – is not one of the agencies represented in the network.

8.3.3
Central-Local Relationships and Capacities

The CRGT was essentially a municipal initiative and took on a bold leadership role. The central government partially devolved authority, and was supportive of the CRGT's efforts, but there was never full recognition of the CRGT's authority to manage the basin. Since 1998 the central government has neither pushed the devolution forward nor terminated the commission. It has kept the commission alive while rethinking and shifting focus concerning environmental and natural resource policy aspects. Thus, neither a complete handover nor a complete abandonment has resulted. This situation is related to the confused legal status of the CRGT, which has to a large extent left it in organizational limbo. Financial resources for the basin management effort have thus always been limited, and the CRGT has never had its own revenue stream. This has severely limited the commission's ability to evolve into something more than a meeting place.

Cantons and municipalities do perform a number of functions, so there appears to be local-level experience with self-governance and service provision in Costa Rica. However, although the law grants municipalities considerable autonomy, in practice they receive insufficient funding to carry out their responsibilities effectively. This lack of resources and capacity at the local level, with no authority in an intermediate level of governance, reflects the generally high level of centralization in Costa Rica. Nor is local-level government particularly acquainted with the water portfolio, as the Institute of Aqueducts and Sewers (Instituto Costarricense de Acueductos y Alcantarillados; AyA) has increasingly assumed many of the planning and water supply functions of municipalities in recent years.

Discussion of how to strengthen local government and decentralize Costa Rica's system has been under way in earnest for more than 20 years, with various legislative proposals introduced in the national legislature. Recently these efforts have begun to bear fruit, with the establishment of a new Municipal Code in 1998 that, among other things, provides for the direct election of mayors (which occurred for the first time in December 2002), and the approval in 2001 of a constitutional amendment assigning 10 percent of the revenue of the regular budget to the municipalities.

The ability of any basin commission in Costa Rica to develop and implement effective water supply management policies is likely to be hampered by the weak framework of water rights allocation. There is a consensus now that it is necessary to have a legal framework to regulate water, and in 1998 a process was initiated, promoted by diverse sectors, to approve a new general Water Law, which is likely to provide for the decentralization of water administration and the formation of local structures.

The Tárcoles basin commission has existed for more than a decade, which should have given it adequate time for implementation and adaptation. However, the central government's treatment of it changed substantially about halfway through that period, and uncertainty and lack of resources have plagued it since. Thus, the basin commission has been unable to make significant changes to its own internal structure and operations to improve basin management over time.

8.3.4
Basin-Level Institutional Arrangements

As regards structural matters, the misjudgments that occurred when creating the internal structure of the CRGT meant that, once it had lost its central government support and its dynamic initial leadership, its status and composition left it vulnerable to becoming more of a discussion forum than a governing body. Even so, its prime function as a forum for information sharing and communication has waned considerably since its time of peak activity as stakeholders have turned to other sector-based forums (union of municipalities, chamber of industry, chamber of agriculture) for information sharing, which may serve operational purposes, but is not conducive to development of an agenda of basin activities.

Efforts to match the basin boundaries appear to have proceeded fairly well. The real difficulty lies in identifying who is responsible for what in the Tárcoles basin. The prevailing and traditional view that water has to be managed by its uses (drinking, irrigation, hydropower) rather than in an integrated fashion has been reflected and reinforced by Costa Rican laws. There is considerable fragmentation and territorialism among agencies and institutes at the central government level. Likewise, at the local level, there is little interjurisdictional cooperation and coordination among municipalities that have been gaining interest in entering water planning and water service business activities.

Nor has there been adequate recognition of subwatershed communities of interest, given the very different characteristics of the upper, middle, and lower basin areas. One of the principal recommendations being considered by the ministry regarding how to proceed and how to restructure the basin commission is the establishment of upper, middle, and lower basin groups (subcommissions) within the overall commission.

8.4
Performance Assessment

8.4.1
Stakeholder Involvement

A number of national-level organizations, mostly governmental, have an interest in water resource management in the Tárcoles (and other) basins. Most relevant of these is the Ministry of Environment and Energy (MINAE), which is the central government agency with principal responsibility for environmental and natural resource management. The transfer of water resources to its portfolio in 1996 was not followed by any significant decentralization of water management responsibilities.

This tendency towards centralization of responsibilities at national level is illustrated by other government agencies. The Institute of Aqueducts and Sewers (AyA) was created in 1961 with authority to design, finance, build, and operate water supply systems created after that date, and is responsible for water supply in the San José metropolitan area, as well as acting as the regulatory authority for water supply and sewerage systems in Costa Rica. AyA is also one of the major polluters in the Tárcoles basin because of the lack of treatment for its sewerage system, which discharges to the surface waters of the basin. Other governmental organizations with water resources as part of their portfolio, or who regulate or support other basin stakeholders, include the Ministry of Health, the Ministry of Agriculture and Livestock, and the Public Service Regulatory Agency.

An important role is played by the Costa Rican Electricity Institute (Instituto Costarricense de Electricidad; ICE) and its subsidiary the National Power and Light Company (Compañía Nacional de Fuerza y Luz; CNFL). ICE, an institute with budgetary and functional autonomy, was created in 1949 to plan and conduct the rational development of energy for the country, especially from water resources. ICE and the CNFL have constructed a series of reservoirs for hydropower generation in various rivers in the Tárcoles basin. In addition, since 1992 Costa Rican law has allowed private companies to obtain permits and concessions to produce and sell electricity; the Costa Rican Association of Energy Producers (Asociación Costarricense de Productores de Energía; ACOPE) represents their interests.

The National System of Conservation Areas (Sistema Nacional de Areas de Conservación; SINAC) is the agency charged with the management of ecologically significant and vulnerable protected areas, many of which are located around the periphery of the Tárcoles basin, mainly in the upstream basin area. There are also several nongovernmental organizations oriented toward natural resource protection, sustainable urban development, or both. Some have had representatives on the Tárcoles basin commission. They include the Center for Environmental Law and Natural Resources (Centro de Direcho Ambiental y de Los Recursos Naturales; CEDERENA), FUDEU, and the Federation of Environmental Groups (Federación Costarricense de Conservación del Ambiente; FECON).

Municipalities are responsible for operating the water supply systems under their authority that existed prior to the creation of AyA, as long as they maintain a minimum level of quality and efficient service and obtain a concession from the national government for their water use. There has been a tendency toward the centralization of the service in AyA, because of the poor service provided by the municipalities.

Finally, industrial businesses (especially manufacturing and food and beverage processing concerns) and agriculturalists (irrigated agriculture is a major water use sector) are noteworthy stakeholders in the Tárcoles basin.

A large number of stakeholders, therefore, have an interest in efficient management and utilization of water resources in the Tárcoles basin, as might be expected in an area with such a degree of economic and urban development, However, a number of factors limit the opportunities for meaningful stakeholder involvement at basin level; these include the reluctance of national government to devolve responsibilities for management of resources, including water; the decline in the effectiveness of the CRGT; and the lack of further institutional development since the establishment of the CRGT in 1992.

8.4.2
Developing Institutions for Integrated Water Resource Management

One of the most intriguing questions about the Tárcoles case is what motivated the creation of the basin commission in the first place. The municipality of San José, and one official therein, Hubert Mendez, took the lead in convening the initial meetings and workshops that led to the formation of the CRGT. Three elements of motivation seem to have spurred this process (Ballestero 2003): the commitment of Mr Mendez to environmentally sustainable urban development, including the restoration and maintenance of urban rivers; the concern of officials in the San José area for quality-of-life issues, including water quality and water availability, once the population of the metropolitan area had passed 1 million; and the realization that potential solutions to these problems involved reaching beyond the borders of the municipality of San José. Creating a cross-jurisdictional entity such as a river basin commission was a means of raising awareness and influencing actions in neighboring jurisdictions and in the private sector and civil society.

The other jurisdictions and private sector or civil society organizations that participated in the formation and early years of the CRGT saw advantage in a system of water resource management that would allow sustainable growth, and sought representation in a reform process that would work towards that end. Without a basin commission, local government and private sector or civil society organizations would more likely be on the receiving end of central government policy rather than helping to shape it (as became evident in the subsequent period of CRGT decline).

Crucial to the early success of the CRGT was the participation and support of AyA and ICE, the two largest stakeholders in the Tárcoles basin in terms of water use and policymaking. Both have been represented on and participated regularly in the Tárcoles commission. AyA was keen to support a process that might lead to measures such as penalties for sewage dumping or subsidies for sewage treatment. It also saw involvement with the CRGT as the best way to protect its own management attributions, which helps explain why it did not object strongly when the CRGT declined and was folded into MINAE. ICE and the CNFL also sought to protect their own positions through participation rather than obstructionism; they opposed proposals by the private hydropower producers, represented by ACOPE, that ICE and the CNFL be brought into the system of water use concessions and tariffs, arguing that concessions would negatively impact investments and tariffs would increase costs to customers.

It was, however, the policy considerations of MINAE that proved most crucial in shaping the institutional development of the CRGT. Although the support of MINAE was a major stimulus to the early advances made by the CRGT, it subsequently failed to resist the temptation to maintain control of the water resource pillar of national environmental policy, for at least two discernible reasons. One was the obvious bureaucratic reason of seeking to maintain control over an important policy topic and the governmental resources devoted to it. The other was more subtle; although at least some MINAE officials accepted that integrated water resource management on some sort of regional scale (river basin or otherwise) was desirable, they preferred to develop and implement such policies and practices for the nation as a whole rather than piecemeal. The Tárcoles basin in particular was considered much too important for MINAE to leave autonomous, and the lack of a strong legislative foundation for the CRGT made it easier for MINAE to assert its control over the organization.

8.4.3
Effectiveness and Sustainability

As already noted in Sect. 8.2.3, the CRGT was for a period in the 1990s able to initiate and lead important basin improvement activities. Agribusiness contamination of water, especially from coffee processing operations, was reduced through the Voluntary Plan program. Although forestland and farmland are still being lost to urbanization, aggravating erosion, flooding, runoff, and contamination problems, reforestation efforts championed by the basin commission certainly helped slow the degradation by as many as 150 000 trees.

Subsequent changes of leadership at the CRGT and its changed relationship with MINAE – a change that resulted in more central government control but less central government support – are associated with a decline of CRGT activity, visibility, and stakeholder participation. A number of basin management issues remain unaddressed and unresolved in the aftermath of that change.

The Tárcoles basin still lacks sewage treatment, and river water quality conditions therefore continue to worsen as the basin population grows. Septic tanks used by many households and businesses in the basin are not being replaced with a sewage collection and treatment system, and groundwater quality is increasingly threatened as a result of septic systems as well as agricultural and industrial water and land uses. Industrial waste treatment occurs in some locations in the basin, but coverage remains incomplete.

The water rights system in Costa Rica inhibits effective demand management. The current concession system does not cover groundwater use, or surface water use by public hydroelectric suppliers. ICE, as noted, has been outside the surface water concession system despite being the largest surface water user. This is explained on the basis of ICE's status as a state institute, but also on the grounds that hydropower use is nonconsumptive. Both rationales are defensible, but the exemption creates political difficulty in getting other surface water users to accept registration, limitations, and tariffs on their water use. The tariff system for agricultural water use continues to impose fees based on cultivated area rather than metered water use, providing no economic incentive in favor of efficient water use. Furthermore, groundwater use appears to be subject to no control whatsoever, though there is evidence of overdraft in the San José area. These national policy problems delay improvements to basin conditions, exacerbating the problems arising from the tenuous status and institutional position of the CRGT.

Finally, though, it needs to be pointed out that the Tárcoles basin experience has led to greater participation of a number of segments of society in water-related issues. This in itself is an achievement in a traditionally centralized country. It has also had a certain influence on the strides made towards the new Costa Rican Water Law and towards basin management approaches in other basins.

8.5
Summary and Conclusions

8.5.1
Review of Basin Management Arrangements

The Tárcoles basin case provides a useful example of the vulnerabilities of bottom-up initiatives for basin management. Such initiatives often lack a well-defined legal role and mandate. They may be dependent upon higher levels of government for funding and technical support, and thus become vulnerable to political changes that shift governmental control and policy direction. They also may lack the authority to undertake cross-boundary efforts to resolve basinwide problems and conflicts. The CRGT's experience exhibited all of these characteristics.

More specifically, the following important features are associated with the Tárcoles case:

- The start-up of the commission in a bottom-up format initiated by some of the large stakeholders was initially very successful and quickly showed a number of results, indicating the possibilities for basin management. Nevertheless, it was heavily reliant on high-level support.
- The central government's commitment to river basin management generally, and in the Tárcoles basin in particular, has been uneven and inconsistent. This was especially evident with a change in government and is quite usual, especially in developing countries.
- The basin management approach had a strong champion. Once the champion left (and the above-mentioned political changes took place) the still young and fragile institutional setup became stalled and relatively ineffectual.
- The severity of problems in the basin, and the difficulty of marshaling the financial resources to address them, stretch the management challenges beyond the capabilities of local action without sustained commitment of central government or external support.
- Flaws in the basin organization structure and authority kept it from exercising autonomous authority to govern basin management, and diminished the commitment of some important local actors to it.
- The past and current water rights established in the Water Law have been notably unhelpful to integrated water resource management in the Tárcoles or other basins and it is not clear if the Tárcoles experience has helped reshape these laws. However, other basins have learned from the Tárcoles and adopted different approaches.

- The biggest water interests are national scale and have their own agendas. They either must have incentives to participate (which they currently do not have) or the government must act more forcefully if it wants to promote better basin management.
- Pollution may not be perceived as acute a problem as water scarcity in other contexts, so the political pressure to deal with the issue is relatively low.

The current situation in the administration of water resources in Costa Rica remains characterized by fragmentation and dispersion of responsibilities in a large number of institutions, several of which operate on a national scale. At least 15 agencies are involved in local and national water administration. As a result, there are serious problems in the distribution of responsibilities, with overlaps in some areas and vacuums in others. There is no coordination between these institutions, and their systems of administration differ. They were created to fulfill specific functions (such as irrigation, drinking water supply, hydroelectric generation, and sanitation) and lack an outlook that envisions an integrated approach to water resource management.

8.5.2
Future Prospects

Despite the fact that MINAE is responsible for supervising water resources, in 2002 the National Water Council was formed and charged with the "harmonization of water legislation and the coordination of research, uses, development, utilization and conservation of water in the different departments and institutions of the state". The Ministry of Health, an agency with responsibilities for water quality, pollution, and health, was appointed to coordinate this council, which created distortions and overlapping roles for both ministries. The formation of the council was a product of the leadership vacuum created by MINAE as the supervisory agency, was a temporary measure, and was no substitute for an adequate institutional framework for integrated water resource management. Accordingly, the National Water Resources Management Strategy, recently passed at the time of writing, has significantly redefined the water resource sector, empowering MINAE to play a greater leadership role and thereby reducing the role of the Ministry of Health.

There is thus hope for change. There is an ongoing process to reform the legal and institutional water framework in the country and the new national water policy and water resource strategy are first steps in the right direction. The draft Water Law contemplates a vision of decentralization and includes the creation of river basin organizations. However, these would primarily exercise control functions, rather than a more proactive executive function in the basin.

The weaknesses and strengths of the CRGT have become fairly well known, as have its failures and successes. And through the past decade, the different actors have accumulated experience in water resource management that can be applied to any future process of integrated management in the Tárcoles basin.

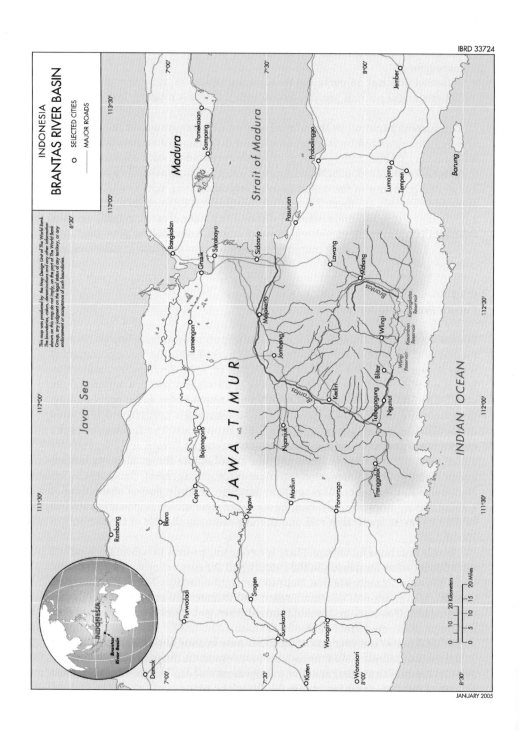

INDONESIA

BRANTAS RIVER BASIN

o SELECTED CITIES

— MAJOR ROADS

This map was produced by the Map Design Unit of The World Bank.
The boundaries, colors, denominations and any other information
shown on this map do not imply, on the part of The World Bank
Group, any judgment on the legal status of any territory, or any
endorsement or acceptance of such boundaries.

Java Sea

Madura

Strait of Madura

JAWA TIMUR

INDIAN OCEAN

Barung

Brantas

Brantas

Brantas

Rembang
Blora
Cepu
Bojonegoro
Ngawi
Lamongan
Gresik
Surabaya
Bangkalan
Sampang
Pamekasan
Sidoarjo
Mojokerto
Jombang
Lawang
Pasuruan
Probolinggo
Malang
Wlingi
Kesamben
Reservoir
Karangkates
Reservoir
Wlingi
Reservoir
Lumajang
Tempeh
Jember
Madiun
Ngawi
Nganjuk
Kediri
Blitar
Tulungagung
Ngunut
Trenggalek
Ponorogo
Purwodadi
Sragen
Surakarta
Wonogiri
Klaten
Wonosari
Demak

INDONESIA
Brantas
River Basin

0   5   10   15   20 Miles
0   10   20 Kilometers

7°00'
7°30'
8°00'
6°30'
8°30'

111°30'
112°00'
112°30'
113°00'
113°30'

Chapter 9

Indonesia: Brantas Basin

A. Bhat · K. Ramu · K. E. Kemper

9.1
Background

9.1.1
Introduction

Indonesia, like many other developing countries pursuing a path of rapid economic growth, has found its water resources coming under increasing pressure, and recognizes the need to develop a program at the basin level to address water quality and scarcity issues during the next 25 years.

The Brantas basin area of East Java is an economically developed region of national importance to Indonesia. Water infrastructure development for purposes of flood control and regularization of supply for irrigation, industry, and power generation has historically been the responsibility of central government line agencies; the establishment in 1990 of the Brantas River Basin Management Corporation (Perum Jasa Tirta I; PJT I) marked a major shift in policy by placing emphasis upon the management aspects of water resources at the river basin level, rather than primarily on water and infrastructure development. How this process is being managed within the context of wider administrative and fiscal decentralization, during a period of rapid economic, political, and institutional change, offers significant lessons to other countries in similar circumstances.

9.1.2
Basin Characteristics

The Brantas basin is located within the province of East Java in Indonesia. It has an area of approximately 11 800 square kilometers, 24.6 percent of East Java's land area. It is bordered by high mountains on several sides. The Brantas River itself is 320 kilometers long, rising in the Arjuno volcanic massif (Ramu 2004). The river is regulated by a number of dams and reservoirs during its course, and at its delta the New Lengkong Barrage divides the Brantas into the Porong, which serves as a flood diversion channel during the rainy season; the Surabaya, which is the primary source of raw water for Surabaya city; and the Porong and Mangetan canals, which provide irrigation water for the paddy-growing region. The river finally discharges into the Madura Strait. The basin contains two active volcanoes – Mount Semeru and Mount Kelud – whose erupted materials contribute to soil fertility but can cause loss of life, property damage, and reservoir sedimentation.

The basin's tropical monsoon climate is characterized by mean annual temperatures of about 25.5 degrees Celsius and annual rainfall averaging 2 000 millimeters, mostly falling in the rainy season from November to April. Large variations in rainfall totals can occur from year to year.

The average surface water potential in the Brantas basin is estimated at 12 billion cubic meters, with the average flow estimated at 3 billion cubic meters, or 25 percent of available surface water. There are over 400 deep wells for irrigation and urban water supply, and over 27 000 shallow wells for rural water supply. The infrastructure in the basin includes eight large reservoirs with a gross storage capacity of about 650 million cubic meters, and other major structures such as barrages, pumping stations, diversion gates, and tunnels valued at US$869 million in 2000 (Ramu 2004).

The Brantas basin has strategic significance for Indonesia and East Java, containing 23 percent of the province's forestland, 56 percent of its arable land, and 43 percent of its irrigated land. The agricultural economy centers on (mostly irrigated) paddy cultivation, and the basin contributed 32 percent of East Java's total rice production in 2000. Other important food and cash crops are maize, cassava, soybeans, peanuts, tobacco, coffee, and sugarcane.

The basin's population is about 15 million, having increased by over 50 percent during the period 1975–2005, and has an average density of 1 249 persons per square kilometer. The basin contains two major cities – Surabaya (population 2.5 million) and Malang (743 000) – and in 2000 contributed 56 percent of East Java's gross domestic product. Although industrial employment has grown considerably, most of the basin's residents are agriculturalists with small to medium-sized landholdings. Irrigators accounted for 76 percent of surface water diversions in 2000, while industrial and domestic users diverted 9 percent and 14 percent respectively (PJT I 2000 data).

Overall, the Brantas river system supplies a great variety and number of domestic, industrial, and agricultural users, all depending on reliable access to sufficient amounts of safe water. In addition, its dynamic socioeconomic development is expected to continue contributing to the growth of East Java and Indonesia, thus its formal and legal designation as "nationally strategic" by the central government.[1] This important feature will be explored further below since it has had a major impact on the institutional and organizational options for management of the basin's water resources.

9.1.3
Water Resource Problems

The intense industrialization, agricultural development, and population growth within the Brantas basin over the past three decades, combined with its climatic and physical features, have resulted in several critical water resource problems, including pollution, flooding, river and reservoir sedimentation, and seasonal water scarcity. While large investments have been made, including construction of critical infrastructure and

---

[1] There are about 15 such basins. Although since 1999 the government of East Java has been wholly responsible for the Brantas basin's water resources, the government of Indonesia is still legally entitled to assume management control of basins defined in this manner and is responsible for their financing.

institutional investments for improved water resource management, significant problems remain to be resolved.

Pollution

The most serious problem currently stems from untreated effluent from industry, domestic users, agriculture, and livestock breeding draining into the Brantas River. Pollution loads are primarily from domestic and industrial sources. Rapid urban growth and the lack of resources to address sanitation, sewage, and solid waste have resulted in an increase of pollution in urban areas. On average, 65 percent of Brantas basin inhabitants are served by public, shared, or private sanitary facilities. Biological oxygen demand is about 10–20 milligrams per liter in the Surabaya River at Surabaya city and 8–15 milligrams per liter in the Brantas River at Malang city, amounts that exceed the assimilative capacity of the rivers during the dry season (Usman 2000).

Industries are required by law to treat effluents, but regulatory institutions are very weak and lack resources to monitor and enforce regulations. While the Clean River Program (Program Kali Bersih, or Prokasih) has attempted to reduce industrial pollution loads, the emphasis on economic development and support of export-related industries has been a disincentive to enforcing regulations on pollution control. Agricultural waste pollution is not as significant a factor given that agricultural activity mainly takes place during the rainy season when the flow of water is sufficient to flush out pollutants. However, agricultural pollution accumulating in reservoirs and rivers during the dry season from irrigation return flows is mobilized during the wet season in some reaches of the basin. In most reservoirs, nutrient depletion is causing eutrophication. Total pollution load in the basin has increased almost threefold in the last 10 years.

Flooding

The Brantas basin experiences flooding in its lower reaches due to low gradients, encroachment on flood plains in rural and urban areas, and sedimentation in the rivers and in the reservoirs. The volcanic activity of Mount Semeru and Mount Kelud adds sediment to the river system, reducing its discharge capacity and increasing the potential for flooding. A further factor is wide-scale deforestation in the upper reaches of the basin to expand agricultural land use, exposing large areas to sheet erosion and again increasing the sediment load of the river system. These factors have also led to a speeding up of the rate of siltation in reservoirs, with detrimental impacts on reservoir infrastructure, water storage, and power generation. The Wlingi, Lodoyo, Sengguru, and Sutami reservoirs have all been adversely affected. Flood control infrastructure has been constructed to provide protection for return periods of 10 to 25 years. Nearly 60 000 hectares of land used to be flooded annually prior to flood control implementation.

Water Shortages

The rainy season provides an abundant water supply for the basin but water availability during the dry season is often barely sufficient to meet existing demand when instream water quality objectives are taken into account. Water supply is particularly affected in the high-consumption region below the New Lengkong Barrage, which includes the delta irrigation system, the Greater Surabaya municipal area, and a high concentration of industries. Sugarcane factory operations, which make up 33 percent of industrial water demand, take place in the dry season, leading to diversion of irrigation supplies to meet industrial demand during low-flow years, which contributes to crop losses.

9.2
Decentralization Process

9.2.1
Pre-reform Arrangements for Water Resource Management

Due to its strategic importance, the Brantas basin has been a subject of the central government's attention for decades. In 1961 the Brantas River Basin Development Project was created, which focused on infrastructure solutions to the water resource management challenges encountered in the basin. The Brantas project continues to exist as an agency to implement water infrastructure and is still managed, funded, and implemented by the central government. The project was set up to implement the Brantas River Basin Flood Control Plan (Master Plan I), which reflected the priority given to control of the regular flooding causing devastation in the basin. The basic concept of the plan was "one river, one plan, one coordinated management". Japanese postwar reparation funds were used to implement large technical developments – dams, flood diversions, retarding basins, and riverbed channels. The national Ministry of Public Works established and oversaw the project. Table 9.1 presents a timeline of Brantas basin management, and illustrates the emphasis on physical infrastructure during the first 30 years of the Brantas project.

9.2.2
Impetus for Reform

In 1990 it was acknowledged that lack of incentives for operation and maintenance was making the investments described in the previous section unsustainable. A different approach was therefore sought, leading to the establishment of Perum Jasa Tirta I (PJT I),[2] a state-owned company for basin management, independent from the Brantas project (see Sect. 9.2.3). The early 1990s therefore ushered in a new era of basin water resource management not only in the Brantas basin but in Indonesia.

---

[2] A *perum* is a corporation that manages both revenue-generating activities that must be self-supporting, and non-revenue-generating public welfare tasks (such as flood control and subsistence irrigated agriculture) that are wholly or partly supported by government. Thus assets such as multipurpose dams and flood control levees are not included in the corporate balance sheet, as a return on these assets is not required. This is in contrast to a *peresero* such as PLN (the Electricity Corporation), which fully owns all its revenue-generating assets, all of which are included in its balance sheet and for which there needs to be a satisfactory financial rate of return.

Table 9.1. Master plans and organizational developments in the Brantas basin

Year	National-level event	Brantas basin-level event	Description
1961		Master Plan I	Emphasized flood control by constructing dams in the upper reaches and river improvement to increase capacity
		Establishment of Brantas River Basin Development Project	Plans and constructs infrastructure for basin under authority of Ministry of Public Works
1969–1998	General Suharto's New Order Government in power		Water resources and other governmental functions consolidated to center
1969–1994	First 25-year development plan (PJP I)		Water resource policy emphasized rice production self-sufficiency
1973		Master Plan II	Emphasized irrigation development to support government policy on rice self-sufficiency
1974	Water Law 11/1974		
1985		Master Plan III	Emphasized water supply for domestic and industrial users to support the government policy on industrialization and urban development
1990		Establishment of PJT I	Public company established for operation and maintenance of infrastructure and water operation in the Brantas basin
1994–2019	Second 25-year development plan (PJP II)		Emphasizes integrated water resource management and operation and maintenance of infrastructure
1998		Master Plan IV	Emphasizes institutional approaches to address conservation, management of basin water resources, and pollution problems
1999	Decentralization Law 22/1999 and Fiscal Equalization Law 25/99; new government regulation on structure of PJT I		Change in relationship of PJT I with respect to the provincial and district governments
2004	New Water Law 7/2004, revised Decentralization Law 32/2004 and Fiscal Equalization Law 33/2004	New Master Plan proposal	Implications for future management? Provides both province and center with a greater say in matters of decentralization and fiscal resource allocation to the districts. Gives province a greater role in PJT I affairs

At the national level, Indonesia's second 25-year development plan (Pembangunan Jangka Panjang; PJP II) commenced in 1994 with an emphasis on integrated development and management of water resources, with greater focus upon the operation and maintenance of infrastructure. This new plan illustrates the shift in mindset of Indonesia's administrators from a single-purpose focus to a multisector river basin approach to promote integrated water resource management. It was decided that authority and responsibility for irrigation management, which had been the primary focus in the previous long-term plan, was to be transferred gradually to the district and provincial levels and was to include farmer participation as part of government policy to increase regional autonomy, while the allocation of water among irrigation and other uses would make up a core function of basin management (Ramu 1999). Indonesia began to set up national policies towards organizing institutions and integrating management functions on the basis of hydrological boundaries rather than administrative boundaries, a step that is vital to efficient management of basin water resources.

9.2.3
Reform Process

As described in the previous section, the start of a new approach to water resource management in the Brantas basin was marked by the establishment, in 1990, of PJT I as a state corporation to primarily operate and maintain the basin's major water infrastructure (excluding irrigation supply systems) and to manage its water resources as a bulk water supplier on behalf of central government, which remains the owner of the infrastructure. As defined by government regulations, PJT I is assigned to plan and operate day-to-day activities, maintain records, undertake minor maintenance, and assume responsibility for operational management. The company can undertake some repair and rehabilitation activities necessary for operational purposes, within its financial limits. Major rehabilitation or additions to water infrastructure to support efficient water management is undertaken by the Brantas project with central financing. Due to the basin's relatively high level of economic development, PJT I can achieve a reasonably high level of operation and maintenance cost recovery from water users: hence the logic of corporatizing the water resource management function with respect to bulk water supply and allocation.

Globally, this outsourcing only of water resources and infrastructure management functions to a freestanding public service company has few parallels.[3] PJT I manages bulk water supply allocation (including for irrigation), water quality, flood control, river environmental management, and water resource infrastructure for 40 rivers, constituting the majority of significant water resources in the basin. The remaining secondary, tertiary, and quaternary rivers are served by the province through basin water resource management units (balai pengelolaan sumber daya air; Balai PSDA) if they cross districts (kabupaten), and by the district offices (dinas offices) if they are within the district boundary. Box 9.1 describes the management and budgetary characteristics of PJT I.

Previously, the Brantas project had been responsible for planning for Master Plans I, II, and III; however, in accordance with ministry regulation 56/1991, which delineates the preparation of basin master plans as a PJT I task, PJT I was involved

---

[3] The Water Resources Management Company (COGERH) of Ceará state in Brazil has been established in a similar manner and with similar objectives (see Chap. 6). Globally this approach is often recommended in order to increase the focus on and incentives for the management function of water resources, but it has so far seldom been implemented.

Box 9.1. Management and Budgetary Characteristics of PJT I

Management of Brantas PJT I is through a supervisory board with operational management under a president director assisted by three directors. Its structure indicates ministerial authority over its affairs, which is typical of Indonesian state corporations. The supervisory board, which is answerable to the Minister of Finance and the Minister of Public Works, carries out the general supervision of the corporation, including implementation of its work plan and annual budget. The board comprises members selected from the Ministry of Finance, the Ministry of Public Works, and agencies whose activities are related to the corporation. The governor of East Java also sits on the board. The power of the supervisory board, as stipulated in the regulation, gives a degree of management autonomy to the basin agency, but the fact that PJT I's infrastructure operation and maintenance costs are subsidized by the Ministry of Public Works, as well as its technical supervision, gives the ministry considerable influence over its operation.

Budgetary arrangements between PJT I and central government are not as straightforward as was intended. PJT I seeks to fund itself through the water it supplies to industry, hydropower units, municipal water suppliers, and other sources such as leasing of land, recreation, and consulting, thereby covering operation and maintenance costs and turning in a set profit to the central government. In reality, an economic downturn in Indonesia has reduced revenues and increased operation and maintenance costs, added to the fact that, as in many other countries, farmers are exempt from payment for irrigation water deliveries, although they constitute the largest water user group. In addition, tariff setting is not in the hands of PJT I but is decided through a political process: the Ministry of Finance sets tariffs for hydropower users and the provincial government sets tariffs for municipal and industrial users. As a result of these factors PJT I is currently able to cover only 30 percent of actual needed operation and maintenance cost of infrastructure under its management, after the pre-established "profit" has been paid to the government. Also, PJT I is subsidized by central government, as the salaries of the staff are paid from national funds.

PJT I operates a resort area along the Selorejo reservoir with the interest of promoting recreation, tourism, and water sports within the basin, and rents out its land for agricultural purposes. For developing such nonwater sources of income, which comprised 25 percent of revenues in 2003, PJT I seems to have some level of autonomy as long as it fulfills its larger organizational objectives.

---

in generating Master Plan IV with the consultation of local government and users. Planning is now recognized by the central government to be part of PJT I's management function. Once the plan is accepted by the central government, PJT I can set up its long-term action plan to implement it. Master Plan IV emphasized conservation and basin water resource management – institutional approaches for proper water governance. PJT I also completed a long-term (1999–2020) plan with assistance from the Japan International Cooperation Agency (JICA), and is in the process of implementing parts of this plan while awaiting finalization of future investments by the government.

9.2.4
Current Situation

There are a number of agencies that are involved fully or partially, directly or indirectly, in water resource-related functions in the Brantas basin. Table 9.2 summarizes the roles

Table 9.2. Water resource management institutions and roles in the Brantas basin

Management level	Institution	Current water management attributions
Federal government	Ministry of Public Works	Manages strategic basin planning and its development, oversees provision of licenses by governor of East Java, management of irrigation systems above 3 000 hectares through comanagement, gives guidance on flood risk assessment, precautions, and disaster management
	Ministries of Finance, Forestry, Mining and Energy, and Environment	Play a regulatory role in their specific areas; many responsibilities devolved to the provincial government and its various sector dinas
	Ministerial Coordination Team (Tim Kordinasi): Expected to be replaced by a National Water Council with stakeholder representation in 2006	Coordination of water resource policy at national level
	National Power Corporation (PLN)	State-owned profit-generating corporation. Owns and operates hydropower plants in the basin, while PJT I operates the dams and provides water for hydropower production. Participates directly in PTPA
Province/district	East Java Water Resource Services Office (Dinas PUP)	Provincial water resource development and management, with focus on activities that transcend district boundaries. Manages irrigation systems between 1 000 and 3 000 hectares. Provides provincial governor with technical assistance in water resource management policy, infrastructure development, operation and management of irrigation facilities, issuing abstraction permits, and related activities
	Provincial and district irrigation commissions (komisi irigasi)	Irrigation management
	Provincial Water Resources Committee (PTPA): Expected to be replaced by a Provincial Water Council with stakeholder representation in 2006	Serves as a coordinative body to provide operational policy direction for Brantas basin water resource development and management
	District Water Resources Council to be set up in 2006 with stakeholder representation	Coordination of water resource activities in the district
	Domestic water supply companies (PDAMs)	State-owned regional water supply enterprises to supply treated drinking water to urban areas. Dinas PUP issues license for raw water extraction, while PJT I is responsible for delivery

of some of the actors at national, provincial and district, and basin levels that have primary or significant roles and responsibilities in the planning, development, operation, management, or regulatory aspects of basin water resources.

Table 9.2. *Continued*

Management level	Institution	Current water management attributions
Basin	Brantas River Basin Management Corporation (PJT I)	Mandate is to manage bulk water quantity, water quality, conservation, and maintenance of water resource infrastructure (including major infrastructure operation and maintenance for 40 rivers)
	Basin water resource management units (Balai PSDAs)	Operate, manage, and maintain infrastructure and water resources in rivers not under jurisdiction of PJT I. There are three Balai PSDAs in the Brantas basin. Manage interdistrict irrigation systems and irrigation systems with 1 000 to 3 000 hectares and act as regulatory arm of Dinas PUP
	Basin Water Resources Committee (PPTPA) to be reconstituted with stakeholders in 2006	Coordination of basin-level management activities and resolution of conflicts
	Water user associations (WUAs)	Manage water distribution for irrigation within tertiary blocks (50–150 hectares in size)
	WUA federations (WUAFs)	Groupings of WUAs in irrigation schemes; participate in management functions with the agencies

For the technical aspects of basin management PJT I solicits the guidance of the Ministry of Public Works, which supervises PJT I's management functions. The Provincial Water Resource Services Office (Dinas PUP) serves as a regulator for PJT I. District-level government provides support for operational matters, providing enabling conditions at the local level for PJT I. The Ministry of Finance sets tariffs for hydropower users. The governor, who serves as the president's representative in the region, sets tariffs for municipal and industrial users, and the minister ultimately proposes the rate by regulation, further signifying central government's continued influence on fiscal aspects of basin management. Thus, PJT I has no control over the tariff of its bulk water supply services and its revenue is controlled by water rates fixed by political and socioeconomic considerations and not cost-plus considerations. Further, irrigation bulk water supply and flood mitigation services derive no revenue, as these services are exempt from paying service charges under the Water Law. This explains, among other reasons, why PJT I cannot meet its full operation and maintenance costs.

9.3
Application of Analytical Framework

This section considers how the previously identified analytical factors deemed to influence the outcomes of decentralization of basin management structures (Chap. 1) have affected the Brantas basin and the performance of its management ystem since the institution of PJT I.

9.3.1
Contextual Factors and Initial Conditions

The steady economic growth during the Suharto regime came to a dramatic halt with the 1997–1998 financial crises, coupled with the collapse of the autocratic regime. Since then the government, with the assistance of the International Monetary Fund (IMF) and other donors, has launched a strategy of policy and institutional reform and has gradually recovered macroeconomic and political stability, but is still limited in its capacity to obtain development funding.

These trends are mirrored in two distinct phases in the macro context of Brantas basin management. First, PJT I was created during the Suharto regime, indicating that the previous government realized the need for a better approach to ensure management, operation, and maintenance of the infrastructure created by the Brantas project, though administration was still dominated by central government. In the second phase greater emphasis was placed on stakeholder involvement in water management activities while the new Autonomy and Fiscal Decentralization Laws put greater pressure on PJT I to be answerable to provincial and district governments in many aspects of basin water management.

It is interesting to note that even under a centralized regime, a certain devolution of management decisionmaking to the basin level, based on comanagement and deconcentration policies, was possible. This provides an example for other countries that have strongly centralized structures, but may be considering more effective basin-level water resource management.

9.3.2
Characteristics of Decentralization Process

Indonesia's 1945 Constitution bestowed strong powers upon the executive branch, giving the president the authority to determine the nature of regional autonomy. During his New Order regime (1969–1998), General Suharto consolidated powers at the center and established a clearly defined hierarchy, with central government setting policies and regulations, provinces undertaking coordination and supervision, and districts responsible for implementation. Much of the development funding and financial controls remained with the center.

The profound political and economic crisis in 1997–1998, and the forced resignation of Suharto in 1998, required an immediate and large-scale multisectoral response. Autonomy Law No. 22/1999 on regional government devolved central government powers and responsibilities directly to district-level governments in many administrative sectors, bypassing provincial government. The difficulty in governance of over 330 autonomous entities resulted in a revision of the Autonomy Law in 2004, by which provincial government was given the oversight responsibility of district administration.

As regards the water sector, the government implemented an ambitious reform program with external assistance from a number of donor agencies, including the World Bank, which in 1998 approved a US$300 million three-tranche Water Sector Adjustment Loan to provide balance of payments assistance in support of reforms in the management of the water resource and irrigation sectors. The completion of the water sector reform in 2005, including the enactment of the new Water Law of 2004 and its gradual implementation, is expected to enable real decentralization of authority in basin water resource management.

Given this political context, it was hardly surprising that decentralization of water resource management was top down. Even so, key water users and stakeholders – for example industries and municipalities – have an interest in a functioning water delivery framework and do tend to maintain payment of the water fees that permit PJT I to function (albeit not sufficiently to cover full operation and maintenance costs of the system).

With regard to the future outlook, the passage of the new Water Law (No. 7/2004) signals central government commitment to continued reform of the water resource sector in accordance with the agreed action plan developed under the World Bank-assisted Water Sector Adjustment Loan. The law was passed after considerable debate on such issues as the extent of irrigated farmer protection, the privatization of certain water resource service functions, creation of water councils with stakeholder participation, and the establishment of water use rights. These issues may yet prove divisive but the fact that dialogue is taking place between a much broader set of actors than has hitherto been the case is encouraging.

9.3.3
Central-Local Relationships and Capacities

Much power still resides with central government ministries for planning, policymaking, and development financing within strategic river basins like the Brantas. As previously explained, PJT I is responsible for operating, maintaining, and managing water resources in the Brantas basin on behalf of the central government, which is the owner of the infrastructure. The supervisory board of PJT I does not have a stakeholder advisory group or any other form of stakeholder institution overseeing its policies and performance (although this may change with the advent of a reformed Basin Water Resources Council, which includes stakeholders). While this depicts a deconcentration rather than decentralization of central government activities to a basin-level institution to serve national-level objectives and local-level interests, orchestrated by a provincial-level basin agency (Balai PSDA), it must be remembered that in the context of a unitary country holding objectives of fiscal recovery and political stability decentralization is undertaken more carefully.

Prior to the 1999 decentralization reforms, there was little stakeholder involvement or coordination in decisionmaking, implementation, or monitoring within the water sector. The relative influence of stakeholders in the Brantas basin was, and to some extent still is, related to policy direction. Irrigation farmers were the central focus of Master Plans I and II as the country worked towards the national goal of rice self-sufficiency (achieved in 1984), whereas Master Plans III and IV focused on urban and industrial development and integrated water resource management. Industrial interests retain center stage in the post-fiscal crisis era, where economic recovery has been an imperative. Much experience is still to be gained to effectively involve stakeholders, raise funds to cover costs, horizontally integrate water resource management activities, and undertake planning and policymaking at the basin level.

Decentralization reforms in Indonesia are continuing to be customized according to ongoing lessons learned. Time and experience have certainly contributed to PJT I's legitimacy as an operator within the Brantas basin. PJT I has developed strong relationships with various stakeholder groups, universities, and nongovernmental organizations and is well trusted to manage water supply and flood control issues. While there is still considerable ground to cover regarding broader and more direct stakeholder involvement, determining the appropriate level of decisionmaking for basin water management activities, and coordination and integration across agencies, there is a general sense of expectation and enthusiasm among stakeholders regarding the direction toward which institutional and policy changes are heading. There also appears to be a realistic understanding that implementing decentralization is a lengthy and iterative process.

9.3.4
Basin-Level Institutional Arrangements

PJT I is a notable organization in a river basin management context because it is a state-owned company with no infrastructure development responsibility, but with clearly delineated management responsibilities and a limited profit motive. This construct has permitted the company to (a) focus on the river basin as the management unit; and (b) focus on management rather than development and construction. PJT I has thus focused on its basic mandate of good operation and maintenance, creating an effective institutional and organizational base in these areas rather than expanding too soon into more contentious areas such as erosion and pollution. This setup has endowed the company with credibility and reassured water users that the funds it receives from them will be reapplied in the basin for operation and maintenance, an important condition to ensure stakeholders' willingness to financially contribute to basin management expenses.

An additional factor in the success of PJT I has been a dedicated staff and a succession of corporate heads who believed in a transparent and well-managed company and who used strong communication skills to build PJT I into a respected, well-functioning institution. The management efforts to create public awareness for better utilization of the basin waters through appropriate campaigns and involvement of nongovernmental organizations has contributed to improved stakeholder trust in PJT I.

PJT I has further strengthened its technical and managerial capacity through collaboration with foreign firms and local universities in human resource development and establishment of real-time monitoring systems and appropriate management tools. It has also developed a consulting arm to help other basins improve their basin management. It recruits talented staff, provides good benefits, and offers opportunities for staff to improve their abilities through training and participation in national and international seminars.

There are, however, certain areas where weaknesses in the overall institutional arrangements prevent PJT I from functioning effectively. Its mandate is lacking or unclear with respect to a number of issues, including the major recent water resource management challenges of pollution and deforestation; monitoring of effluent discharges and implementation of an effluent discharge fee; and operation of reservoirs for low-flow augmentation to improve instream water quality for municipal water supply while adjusting or reducing deliveries to other water users, such as public irrigation schemes. PJT I is also constrained by the fact that it has to depend on central subsidies both for water supply to irrigation systems and for flood management, which are by law non-revenue-generating functions. With the latest round of institutional reforms to be implemented this situation may change, though it indicates the lack of responsiveness that may occur in a basically centralized system.

9.4
Performance Assessment

9.4.1
Stakeholder Involvement

PJT I has proactively developed strong informal working relationships with many of its stakeholders in order to effectively coordinate activities among central-level ministries, province-level agencies, district-level governments, local water users, and concerned public and nongovernmental organizations. As a champion for basin management, it has embodied the notion of the "Brantas spirit", taking the initiative to conduct public outreach and public relations activities to educate different sectors about the value of integrated water resource management. PJT I staff and management display clear pride in working for the company and in doing a good job.

Within the Brantas basin, PJT I works with a number of local nongovernmental organizations and institutions towards public awareness objectives, for example providing input into environmental issues for students at the University of Brawijaya in Malang. The coordinators of these programs believe that behavior-changing education about water resource issues can reach family and the wider community through the students. As part of its public relations PJT I maintains a botanical garden in the basin and a public garden in the city of Surabaya next to the river to showcase improved river reaches for recreational purposes. PJT I is active in disseminating information concerning water resource issues through brochures, participation in exhibitions, local TV and radio, and university seminars. It maintains a website with real-time information about pollution levels and water levels assessed at various locations in the basin, obtained through its online monitoring activities. Local government officials have been particularly targeted for awareness raising since the initiation of decentralization reforms.

The PTPA is a public accountability mechanism through which PJT I obtains concurrence for its seasonal water allocation plan and reports water balance information to governmental entities representing stakeholders, though for the most part it acts as a forum to discuss water allocation decisions and to some extent conflict resolution. In the new Water Law, representation on the PTPA (the provincial council) and PPTPA (the basin council) is to be balanced between the government and nongovernmental agencies, and they will have a broader role than water allocation. These committees will become coordination bodies where decisions on management policies (planning, implementing, supervising, controlling, and funding) are to be made.

Generally, given the limitations in financial and policymaking autonomy and overall authority in areas such as water quality and watershed management, PJT I is considered to be successful by stakeholders because it is committed to upholding a professional and neutral profile, focusing on being a reliable and accessible service provider for tasks it has most authority over to invoke legitimacy: water allocation and supply, and flood control.

9.4.2
Developing Institutions for Integrated Water Resource Management

The heritage of decades of authoritarian government structure, along with the absence of a uniform national-level water resource policy to assist the navigation among discrepancies in legislation, have contributed to a lack of general coordination among regulators, providers, and users at basin, provincial, and central government levels, leading to overlapping functions and conflicting objectives among agencies and challenging the larger achievement of integrated water resource management objectives. Much of this is exacerbated by the lack of proper representation by nongovernmental stakeholders (water users, industry representatives, nongovernmental organizations) on decisionmaking forums, such as the provincial and basin water management committees.

There are, however, positive signs of changes in water resource management structures that reflect the increasing democratization of Indonesian society. Following the 2004 Water Law, the make-up of coordinating committees at all levels (national, provincial, basin, and district) is being reformulated to formally extend direct participation to stakeholders and interest groups. The National Water Council will replace the Tim Kordinasi, the ministerial team currently responsible for coordinating water resource policymaking at the national level. The council will manage a coordination framework for national water resources, with responsibility for guidance in policy formulation, resource allocation, program implementation, regulatory control, intersectoral coordination, and issue resolution. It will comprise various ministers and stakeholder representatives, playing an important part in presenting an integrated approach and commitment to water resource management at the national level. Similarly, the provincial and basin-level coordination councils will be reformed to strengthen representation from nongovernmental stakeholders.

The interaction of institutions operating at subnational and basin level may be exemplified with reference to water use rights, allocation, and conflict resolution. Indonesia's water rights system involves water use rights, with no permission necessary for basic daily needs, domestic purposes, and livestock. The 2004 Water Law further classifies public irrigation systems for farm holdings below 2 hectares as a basic need. For nonbasic needs, the priority of water allocation is left to regional governments according to basin requirements, though the law does mandate the establishment of a framework for water use rights with domestic needs and existing public irrigation ranked as the highest priorities.

Water licensing was formally established in the Brantas basin in 1991, and involves a process that takes three months to complete – a water user requests a license, PJT I completes a technical assessment reconciling the requested quantity and location of the demands with predicted water supply and availability, the district and PJT I provide a technical recommendation to the governor, who then awards the license to the user. The new Water Law stipulates that requests will go directly to the PPTPA before going to the governor, in order to shorten the process and permit stakeholder involvement at basin level. The fear is that the water use right might lose its flexibility by being increasingly accommodated into the legal and administrative framework, with the added concern of transferability issues, given that the law does not permit development of a water market through direct transferability of water use rights. This may lead to loss of control of water resources to local and foreign interests through privatization.

The PTPA, made up of 80 percent governmental representatives and 20 percent nongovernmental representatives (though this allocation will be changed to provide for a balanced stakeholder representation), serves as a coordinative body to provide operational policy direction for Brantas basin water resource development and management. It meets twice a year – before the rainy season and before the dry season – to decide upon water allocation among various users and the rule curve for reservoir operation.

Conflict of interest among stakeholders exists, particularly during the dry season, when there is not enough water available to cover all sector water demands. Irrigation water users, the largest water consumers, receive only 60–80 percent of their water demand, and their allocation is the first to be reduced. Institutional developments are under way to organize farmer interests through federations of water user associations and to have farmers participate more directly in decisionmaking at the basin level through a newly formulated PPTPA structure and the various district irrigation commissions. The Dinas PUP normally handles conflicts among users in the irrigation system (for example upstream-downstream conflicts) through a negotiation process. Intersectoral conflicts concerning water allocation among other stakeholders are handled by PJT I within the PTPA.

9.4.3
Effectiveness and Sustainability

The performance of PJT I in achieving its objectives is best evaluated by considering the overall policy objectives involved in its development and implementation, as well as the most critical issues of the basin. The main tasks PJT I is mandated to carry out are:

- Operation and maintenance of water resource infrastructure, including sediment removal and monitoring; providing technical recommendations; and preparing land use plans
- Water supply services
- Management of the basin, including water resource conservation, development, and utilization; flood warning and control; preparation of water allocation plans; water quality monitoring; and provision of technical recommendations for water licensing
- Rehabilitation of water resource infrastructure

These activities indicate the broadening scope of institutional development and water resource management in Indonesia. PJT I has achieved results in implementing a reasonably good system of water allocation and management and a reliable flood forecasting system, as well as maintaining major infrastructure in fairly good condition. Managing water quality, catchment conditions, and the river environment, however, are the responsibility of many entities, and there is need for greater institutional coordination and authority to address these issues.

Some of the responsibilities of PJT I merit further comment. As regards operation and maintenance of infrastructure, it is generally difficult to assess the effectiveness of such "soft" nonstructural water resource management functions, while planning, financing, and construction of technical projects have direct and visible outcomes. This can lead to expenditure bias in favor of construction of infrastructure projects and against operation and maintenance if the two functions are housed within the same agency, a bias that PJT I was developed to address. In its circumscribed role of planning operations, undertaking day-to-day operation, maintaining records, carrying out minor maintenance, addressing conflicts, and taking responsibility for all operational management, PJT I is successful. However, since it does not collect sufficient funds to cover operation and maintenance costs, it relies upon central government to cover flood control costs, irrigation bulk water supply costs, and major structure rehabilitation as a social welfare activity. It also relies heavily upon central government, through the Brantas project, to rehabilitate gradually deteriorating infrastructure. It is worth noting, however, that despite the inability of PJT I to cover costs, the structure of user charges as reflecting multiple use interests provides a strong base for eventual full cost recovery.

With respect to resource endowments, industry, hydropower, and municipal water suppliers earn far higher revenues than the agricultural sector and pay service fees in exchange for a regular water supply for their operations. It seems to be generally understood among licensed stakeholders that they are taking on disproportionately large costs and subsidizing basin management as a social duty because the typically small-scale irrigation farmers might not be able to afford to pay water service fees, and because obtaining fees from them is difficult given monitoring and coordination costs.

The new Water Law provides for collection of a basin water resource management fee (for which a government regulation has yet to be issued) to pay for the planning, operation and maintenance, and management administration costs of basin water resource management. Smallholder agriculturalists (with landholdings of less than 2 hectares) in public irrigation schemes, who are major consumers of basin waters, are exempted. For other water users, the fee will be determined by the volume of use and will take into account their economic capacity to afford the payment. In principle, this new fee would become an additional source of revenue from nonsmallholder irrigators and other customers of PJT I and enable it to meet more of its operation and maintenance expenses. This, however, depends on a number of factors, including whether the various levels of government agree that this fee may be added to the existing tariffs, and whether the proceeds are included in the balance sheet surplus to be transferred to the Ministry of Finance.

As regards flood control PJT I plays a primary role, having responsibility for operation and maintenance of flood protection infrastructure and for the flood forecasting and warning system. It coordinates the activities of all relevant agencies and the governor, providing constantly updated information on water levels. Agreements exist between the province and the districts regarding how to manage floodwaters. Flood management can be seen as one of the achievements of the past decade's emphasis on institutional change in the basin. The cost of flood control is partly subsidized by the center.

To contend with sedimentation problems that originate from the basin's active volcanoes and erosion due to deforestation, PJT I uses both dredging and flushing techniques, and participates in national reforestation programs. However, rapid deforestation due to timber harvesting, and uncontrolled upland agricultural development using inappropriate practices, continue to cause serious erosion. Decentralization has led to some division of forest management, with central government issuing large concessions, the province issuing intermediate-sized concessions, and local government issuing small concessions. As such, there is little incentive for local government to manage forest resources well (World Bank 2003). Conservation efforts upstream have not been successful since they involve small plots under local government jurisdiction that have no catchment-wide impact.

For water pollution control, final responsibility lies with the governor, who may delegate responsibility to the head of the Provincial Environmental Pollution Control Office. This agency coordinates all other agencies dealing with water pollution control, which has become increasingly decentralized to provincial, district, and municipal levels. However, the process leading to prosecution of a polluter is unclear, and penalties are perceived to be weak. Some progress is being made towards applying the polluter pays principle, with the national government developing legislation for regional and local government wastewater disposal licensing and fee collection for all river basins. It is likely that this will be piloted in the Brantas basin, with PJT I playing a major operational role. PJT I is currently constructing a water quality monitoring system based at a central station in Lengkong Mojokerto, and has built a laboratory in Malang that is awaiting certification, upon which it will become a more effective instrument in successful prosecution of pollution offenders. Currently PJT I meets the cost of water quality monitoring from revenue obtained from water supply, though a government regulation has now been introduced permitting collection of effluent charges to meet the costs of water quality monitoring and management.

9.5
Summary and Conclusions

9.5.1
Review of Basin Management Arrangements

The Brantas case demonstrates a number of important features that are very distinctive yet provide interesting generic insights for river basin management. These include:

- A proactive central government using a top-down approach to undertake decentralization reforms, arising out of the recognition that water resource management activities are best undertaken at the basin level in order to achieve sustainable results.
- Creation of a state-owned semi-profit-making corporation (perum) as an operator with clear objectives related to management rather than to water infrastructure development; with a motive to balance revenue and expenditure while providing an attainable fixed return on limited commercial activity assets (to be paid to the central government); and with a subsidy for operation and maintenance of its hydraulic infrastructure providing public goods such as flood control and subsistence irrigated agriculture.
- A succession of champions within PJT I who have promoted its basin management approach, supported by a dedicated and well-trained team.
- Within these innovative institutional arrangements, a number of water management instruments are in place and actively being implemented: annual water allocation, based on a functioning monitoring system; existence of limited water use rights in the form of licensing of commercial water uses; financial instruments such as volumetric water use fees; a well-established instream water quality monitoring system; and a functioning flood warning system aiding effective flood management. However, no instruments for water pollution control issues are in place.
- Continued external donor financial and institutional development assistance combined with national resources through the central government, which can afford the costs and expenditures involved (that is, no full financial autonomy for PJT I). However, dependence on central subsidies remains a constraint on the financial viability of PJT I.
- With changing economic and environmental conditions in the basin, there is insufficient authority in PJT I to manage and coordinate broader integrated water resource management issues such as water quality and watershed management at the basin level. PJT I is, nonetheless, successful with respect to tasks it is most directly responsible for. It uses the legitimacy it has gained through successful management to coordinate institutions in areas it has less control over.
- Due to the novelty of the decentralization process, which is implemented in an iterative manner, there is still confusion regarding the relationship between many central, provincial, and local government actors. Also, many new coordinating bodies have been created, but there is as yet little clarification concerning their roles and authority. Pressing issues such as water quality and catchment management suffer from this problem of fragmentation of authority without clear coordination. The new Water Law provides clarification and guidance regarding roles and responsibilities but lower-level legal instruments are required before the Water Law can be implemented.
- Structure of representation is an important issue. Since stakeholders are, to date, represented by governmental agencies in water resource decisionmaking, their interests are not directly voiced. This is expected to change as participatory coordination units, involving nongovernmental stakeholders, are developed at the national, provincial, and basin levels.

9.5.2
Future Prospects

Overall, water resource management in the Brantas basin is on a positive track. The management level in the basin compares favorably with many other river basins worldwide even though much still remains to be done and challenges keep growing. One of the ways chosen by the government to further improve basin management is to drive decentralization forward and to more actively involve key stakeholders, including at the subnational level (province and districts) and actual water users (rather than their representatives at the sectoral government level). At the same time, however, the government continues to consider the Brantas a strategic basin and is retaining its overall powers as well as providing resources. The sector reform under way will enable the basin corporation to become self-financing to a large degree.

In practice, the decentralization process has been gradual and still largely reflects top-down arrangements: central government as policymaker with an executing agency as the implementer, and with local government in an intermediate position. Decentralization in Indonesia has focused on devolving authority directly to district-level actors and this has created some confusion concerning relationships among and degrees of authority between the many central government, provincial, and local actors with overlapping responsibilities. There is a move to provide the provinces with more authority in the decentralized framework as the revised Autonomy and Fiscal Decentralization Laws of 2004 are implemented.

Current reform activities are ongoing and weaknesses and strengths are being assessed with respect to the Indonesian context. The new institutional developments promise greater national-level coordination, clearer province-, district-, and basin-level jurisdictional relationships, and expanded stakeholder involvement at all levels. Experience is accumulating through the development of different forms of basin institutional arrangements. Though there may be disagreement among central government actors as to how and how quickly decentralization should be rolled out, there is growing enthusiasm on the part of provincial and local government actors and civil society for further reform of water resource management institutions. A key challenge in this process will be to develop the capacity, both fiscal and managerial, of the regional governments and to bring in stakeholders in an effective and productive way to tackle the major problems that the basin still faces. Participation by itself will not solve these issues, but needs to be targeted and provide stakeholders with clear incentives to work towards solutions.

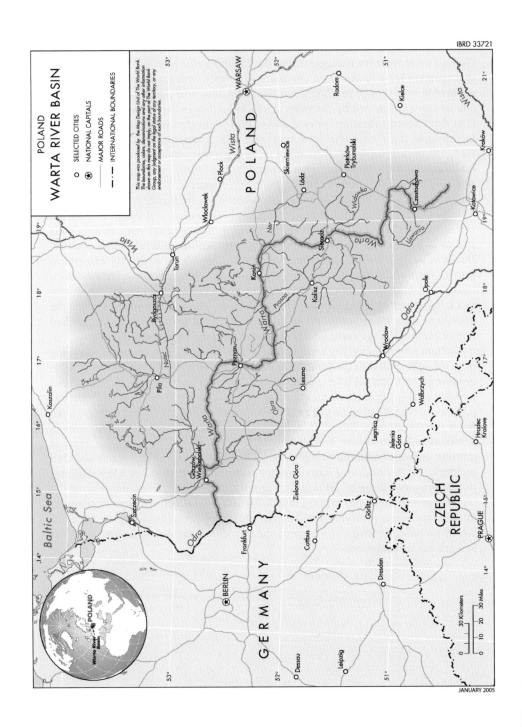

IBRD 33721

POLAND
WARTA RIVER BASIN

○   SELECTED CITIES
⊛   NATIONAL CAPITALS
──   MAJOR ROADS
─ ∙ ─   INTERNATIONAL BOUNDARIES

This map was produced by the Map Design Unit of The World Bank.
The boundaries, colors, denominations and any other information
shown on this map do not imply, on the part of The World Bank
Group, any judgment on the legal status of any territory, or any
endorsement or acceptance of such boundaries.

Baltic Sea

GERMANY

POLAND

CZECH
REPUBLIC

WARSAW
Radom
Kielce
Kraków
Katowice
Opole
Płock
Skierniewice
Łódź
Piotrków Trybunalski
Częstochowa
Włocławek
Toruń
Bydgoszcz
Koszalin
Piła
Konin
Kalisz
Sieradz
Wrocław
Leszno
Wałbrzych
Legnica
Jelenia Góra
Zielona Góra
Gorzów
Wielkopolski
Szczecin
Poznań
Frankfurt
Cottbus
Görlitz
Hradec
Králové
PRAGUE
BERLIN
Dresden
Dessau
Leipzig

Wisła
Wisła
Warta
Warta
Warta
Noteć
Noteć
Odra
Odra
Odra
Obra
Obra
Prosna
Ner
Widawka
Liswarta
Drawa

POLAND
Warta River
Basin
Berlin

0   10   20   30 Kilometers
0   30 Miles

53°   52°   51°
19°   18°   17°   16°   15°   14°
53°   52°   51°
21°   19°   18°   17°   16°   15°   14°

JANUARY 2005

Chapter 10

Poland: Warta Basin

W. Blomquist · A. Tonderski · A. Dinar

10.1
Background

10.1.1
Introduction

The Warta River is a tributary of the Oder, in western Poland, and its basin occupies one sixth of the country. The main water resource issue in the basin is the decline in water quality due to human factors; availability and reliability of supply, and flood control, are other concerns.

Under the Soviet regime following World War II, water resource planning and management were highly centralized at national level, and focused on the construction of physical infrastructure for industrial and agricultural uses. Domestic water supply and sanitation, on the other hand, were decentralized to local level, leading to severe lack of coordination and integrated planning. The process of democratization in Poland during the period 1989–1991 was accompanied by reassessment and reform of governance structures and procedures that gave rise, in the water sector, to the creation of regional boards of water management (later restructured into regional water management authorities), including the one for the Warta basin. However, delays in the reform and updating of national water policy and law led to a gap between national-level guidance and basin-level coordination. This has slowed progress towards integrated water resource management, though commitment to the reform process remains strong.

10.1.2
Basin Characteristics

The Warta River is the largest tributary of the Oder, which forms part of the boundary between Poland and Germany, and is Poland's third largest river after the Oder and the Vistula. The river flows northwards from its headwaters in the mountains of southern Poland, then westwards to join the Oder, and is 808 kilometers in length with approximately 735 kilometers navigable. Major tributaries of the Warta include the Noteć, Prosna, Drawa, Obra, Gwda, Ner, and Welna Rivers.

The Warta basin covers approximately 55 000 square kilometers and is divisible into three major subbasins: the upper Warta subbasin (including the Prosna River watershed), which covers about 21 000 square kilometers; the middle and lower Warta subbasin (to the river mouth at the confluence with the Oder), of about 17 000 square kilometers; and the upper and lower Noteć subbasin, also of about 17 000 square kilometers. The basin contains numerous lakes and reservoirs.

Land use in the basin is 70 percent agriculture and forestry, 30 percent urban and industrial. The basin's population is about 6 770 000, over 34 percent of whom live in cities. By far the largest city in the region is Poznań, the capital city of Wielkopolskie voivodeship,[1] with a population of approximately 600 000. Within Poznań's boundaries alone the river runs for 20 kilometers. Although the city of Łódź has 800 000 residents, it is not entirely within the Warta basin.

10.1.3
Water Resource Problems

The largest water resource management challenge in the Warta basin has for some time been water quality impairment resulting primarily from human uses. There are also issues of water supply availability and reliability, and of flood control. These management issues are linked closely with water uses, of which the largest categories in the basin are industrial (75 percent), municipal (17 percent), and agricultural (7 percent).

Industrial use is primarily of surface water, and agriculture uses a combination of surface water and groundwater. Nearly all municipalities in the basin rely on groundwater resources for potable uses because it is generally of better quality than surface water and requires less treatment prior to use. Only the cities of Poznań, Oborniki, and (for the time being) Łódź use surface water for a significant part of their potable use; Łódź is phasing out its reliance on surface water and will soon use only groundwater.

Water Quality

Concentration of industry, agriculture, and urban populations in the Warta basin has contributed to poor surface water quality (Przybylski 1993). Niemczynowicz (1992) reported that in 1990 only 8.7 percent of the surface water in the Oder basin (which for these purposes included the Warta River) met Poland's category 1 standards,[2] suitable for domestic water supply and salmonid fish habitat. Water in another 24.9 percent of the basin met category 2 standards (suitable for other fish species and for recreational uses), and 24.4 percent met category 3 standards (suitable for industrial use). The remaining 42 percent was not suitable for any use.

Łódź contributes the largest total volume of sewage discharge to the river, followed by Poznań. Until 1990, 99.7 percent of sewage discharged from the Łódź metropolitan

---

[1] Voivodeships are regional levels of governmental administration in Poland, headed by voivodes – state governmental bodies. In relation to self-government functions the same regions are usually called provinces. Regional and local government structure is fairly complicated, and is explained more fully in Tonderski and Blomquist 2003.

[2] Evaluation methods for determining surface water quality in Poland still differ from European Union methods, so these categories do not correspond precisely with the European Union categories of A1, A2, and A3. Nor does the Polish system that is still in use differentiate the quality standards of different water bodies based on their actual or intended use. Waters are simply classified based on their worst-performing pollution indicator. Thus, these measures may overstate the water quality problems somewhat, though it is impossible to say by how much.

area was discharged to the Ner River, a tributary of the Warta, with no biological treatment. Some smaller towns have no sewage treatment plants, so their sewage collections are discharged directly to receiving waters (GUS 2002). Even in the towns that have treatment plants, many are insufficient in terms of capacity and level of treatment (particularly with respect to removing biological material). On the other hand, the amount of wastewater discharged from municipal and industrial sources fell during the 1990s, partly due to reduced water use and partly due to the imposition of fees upon dischargers.

Another crucial water quality problem is contamination from nonpoint sources, which can include agricultural as well as industrial pollutants carried from the land surface to receiving waters. Nonpoint pollution is responsible for 60–70 percent of the country's total nitrogen compounds burden, 40 percent of organic contaminants, and 30–40 percent of phosphorus load (Tonderski 1997; GUS 2001). Nonpoint discharges can also seep into the soil and threaten groundwater with contaminants, though in most cases groundwater in Poland still meets the requirements of European Union and World Health Organization standards (Niemczynowicz 1992; Blaszczyk 2002).

Water Supply Availability and Reliability

Measured by water availability per capita, Poland is one of the most water-poor European countries, and precipitation and runoff in the Warta basin are below even the national average (Kundzewicz and Chalupka 1994). As in many parts of the world, precipitation and runoff in Poland vary substantially from one season to another; greatest volumes of surface flow occur in late spring (with April the peak month on average), and the smallest in autumn (with September the lowest month on average). Supplies also fluctuate from year to year, with mean low flows amounting to only 25 percent of average flows. Drought conditions can occur in the Warta basin, most recently in 1991–1992, when a severe drought caused large losses to the economy and ecology of the region (Kundzewicz and Chalupka 1994). Total water resources translate to a little more than 1 000 cubic meters per person per year, far below the 4 560 cubic meters per person mean for Europe as a whole. Fortunately, central Poland (including the Warta basin) is relatively better off than the rest of the country in groundwater resources.

One way to increase the available resources is to store water in reservoirs, which has been a significant aspect of Poland's past approach to water management. The two largest reservoirs on the main stem of the river are the Poraj and Jeziorsko. The Poraj reservoir, located near the headwaters of the Warta, was built to secure water supply for the Czestochowa steelworks and to provide good conditions for recreation around its shores. The Jeziorsko reservoir, completed in 1986, is located closer to the middle of the main stem of the Warta. It was built chiefly for protecting Konin and Poznań against floods, but also provides electricity, some habitat protection,[3] and supplemental water for drought periods (Penczak et al. 1998).

---

[3] In order to secure the habitat for waterbirds, a stable rising level is maintained in the breeding season from 1 April to 20 June.

Flood Control

Jeziorsko reservoir, with a surface area of 42 000 hectares, a total volume of 203 million cubic meters, and an operational volume of 170 million cubic meters, is the principal flood protection structure in the Warta basin. Considerable effort in Polish water management generally, and in the Warta basin specifically, has been directed toward developing structures to prevent flooding and to store water for droughts. The floods of July 1997 unveiled many shortcomings and defects in the flood protection system throughout the country, leading the government to adopt a national program of reconstruction and modernization. Local and regional flood protection and prevention plans have been developed, with their implementation assisted financially by a central government office created in the aftermath of the floods, the Plenipotentiary for Removal of Flood Effects.

10.2
Decentralization Process

10.2.1
Pre-reform Arrangements for Water Resource Management

During the postwar decades of Soviet dominance, Polish governance and policy emphasized central government planning and control. Water resource planning and management exemplified this trend: from 1960 to 1972, the central Institute of Water Management was responsible for water planning and use, and analysis of water resource information. A restructuring in 1972–1973 yielded a central Ministry of Administration, Country Planning, and Environmental Protection, and an Institute of Meteorology and Water Management.

Throughout this postwar period, water management in the Warta basin and Poland generally focused on technical planning and the construction of physical facilities (for drainage, retention, flood protection, and navigation) to support industrial and agricultural development. Water resource expenditures were almost exclusively for waterworks and relied heavily on central government plans and funding. Centrally appointed district directorates of water management (DDWMs) were established beginning in 1964 to construct and operate waterworks – first five, then seven, located on the principal rivers in Poland, with two DDWMs on portions of the Warta River. On the other hand, domestic water supply, sanitation, and wastewater disposal were decentralized, local functions with no meaningful planning and management at a regional or river basin scale. The extent and quality of these services was especially problematic in rural regions.

Until 1991, the main governmental entities responsible for water management at the subnational level were not fitted to river basin boundaries. The DDWMs worked along the main stems of some rivers, but not on basins as a whole. The other regional body was the voivodeship, of which there were 49 in Poland at that time. The voivodes governing these provinces were responsible for rivers and streams that were not being managed by the DDWMs, for irrigation and drainage, and for issuing water permits. In addition, county-level government entities – poviats and starosts – had certain water management attributes, for example related to flood control and response.

10.2.2
Impetus for Reform

By the late 1980s, as the entire governmental system faced a period of crisis and transformation, Polish water resource professionals understood the need to broaden the focus of water policy toward integrated water resource management at river basin level, taking into consideration natural and ecological requirements, as well as public health and safety and economic development. However, the political environment had not been conducive to such radical changes in policy and direction.

All this changed with the democratic transformation of 1989–1991, an "open policy window" that involved a rethinking of government structures and procedures throughout Poland and provided the opportunity for such organizational reform as the creation of basin management authorities (Kingdon 1995). In February 1991, the government announced the creation of a system of regional boards of water management (RBWMs), conforming essentially to river basin boundaries. Their principal purposes were to arrest the further pollution of water supplies, protect drinking water sources, and aid water users and water user organizations in developing and implementing rational water management. Their responsibilities concerned the balance of water supplies and demands, determining the conditions and terms for the use of basin waters, and developing and maintaining a new water information system.

The RBWMs were related directly to the national government's Ministry of Environment. Each RBWM director was an individual[4] charged by the ministry with management of the basin. There was little provision for public participation or water user involvement in RBWM decisionmaking. Also, the DDWMs were kept in place, maintaining their responsibilities for the operation and maintenance of waterworks on their designated river reaches, so in the Warta basin the RBWM covered the entire basin while two DDWMs still operated along portions of the river's main stem.

10.2.3
Reform Process

The dual structure whereby the new RBWMs operated alongside the older DDWMs resulted partly from the fact that the creation of the RBWMs was supposed to be accompanied by a thorough revision of Polish water law and policy. A new national water policy was expected to provide a basis for different systems of decisionmaking about water resource conditions and efforts to improve them – essentially, integrated water resource management with a more participatory structure for decisionmaking. The reforms in water law and policy took much longer than expected, however.

---

[4] Despite the name "regional board of water management", the organizational structure did not really include a board.

Toward the end of the 1990s, efforts to rationalize management functions and shift policy towards integrated water resource management resumed in earnest. Broader reorganization of Polish government in the late 1990s, including the consolidation of the 49 voivodeships into 16, provided an occasion for reconsidering the distribution of responsibilities for water and wastewater, as well as a host of other governmental functions at the subnational level. Poland's movement toward European Union accession (concluded on 1 May 2004) also made it necessary to focus on integrated water resource management in order to begin aligning Polish policy and practice with European Union standards and expectations.

In late 1999, the Ministry of Environment decreed a merger of the DDWMs and RBWMs and their separate operations into seven regional water management authorities (RWMAs) covering the entire country and corresponding primarily, though not precisely, with Poland's principal river basins. The RWMA in Poznań covers the Warta basin from its source to its mouth at the Oder River.

On behalf of the central government, RWMAs perform planning and coordinating functions within river basins, overseeing the actions of voivodeship and local governments and private users for compatibility with basin water management plans, and maintaining specified waterworks and state-owned reservoirs and other facilities. RWMA Poznań owns and operates the regulatory reservoirs in the basin that maintain river flows. Reservoir releases are also a potential means of addressing water shortages. With respect to flood control and protection, the RWMA's function is mainly to serve as a coordination and information center, providing decision support for flood prevention, control, and response.

RWMAs have a legally recognized role in the water use and discharge permitting procedures that are carried out by voivodeship or local (poviat or starost) offices. This allows RWMA staff to be aware of activities and investments in environmental protection and improvement in the basin, and to object to permit applications that the staff conclude will harm basin conditions.

RWMA Poznań also provides opinions and recommendations to provincial and local governments about financing water quality improvement projects, in particular the construction of wastewater treatment facilities. It has been deeply involved in the process of construction of new wastewater treatment plants for the cities of Łódź and Poznań, which contribute much of the point-source pollution in the basin.

RWMA Poznań receives an annual budget allocation from the central government, distributed through the Ministry of Environment. Some of the RWMA's functions in managing state-owned facilities generate fee revenues, but most of that revenue goes directly to the Ministry of Finance. Overall, 99.5 percent of the RWMA's budget comes from the central government.

RWMA Poznań has 322 employees. About half are highly educated professional staff (for example engineers and attorneys), and the others are technical and operations staff. In addition to its main office (in Poznań), the RWMA has three local offices serving basin subareas: one in Poznań, addressing the middle and lower Warta River and the Prosna River; one in Bydgoszcz, addressing the Noteć River; and one in Sieradz, addressing the upper Warta River.

10.2.4
Current Situation

The current arrangements pertaining to water resource management are very much a product of the new Water Law, finally enacted in 2001. It reformed the RWMAs and added a consultative structure for basin stakeholders; each RWMA must establish a regional council of water management (RCWM), composed of water users and representatives of the other governmental units in the basin.[5]

The new Water Law contains a statement of priorities for the nation as a whole, and of priorities among water uses. Public drinking water is the highest priority use. The law further specifies that groundwater resources should be protected and preserved for public drinking water supplies and other uses that require clean water (for example certain industries). All other consumptive uses (industrial, urban nonpotable, and irrigation) are secondary, and their relative priorities depend on basin circumstances. Water use conditions and priorities at the basin scale are to be determined by RWMA directors in basin plans that must meet with the RCWM's approval after wide public consultation.

The new law also requires the establishment of a National Board of Water Management (NBWM), which will be the principal water management entity at the national level, displacing that responsibility to some degree from the Ministry of Environment and the Department of Water Resources within that ministry. Board members are to include the RWMA directors, who will no longer relate directly to the Ministry of Environment but to the NBWM. Among other responsibilities, the NBWM will harmonize RWMA activities and approve their basin plans and progress reports.

The president of the NBWM is to be advised by a 30-person National Council of Water Management (NCWM), first appointed in June 2002, which represents diverse disciplines and constituencies related to water resource management.[6] Pursuant to the new Water Law, the NCWM's activities are to provide advice on matters of water management, flood control, and drought control.

Water policy in Poland also uses financial instruments to provide incentives for water conservation and for water quality protection. The principal financial instruments used currently include fees for discharges of wastewater or withdrawals of surface and underground waters; fines for illegal or excessive waste discharges or water withdrawals; financial assistance in the form of grants or preferential credit for environmental protection and water management; and tax concessions for those providing ecologically advantageous goods and services.

---

[5] These councils are to consist of about 30 persons, with half of them being water users. The others would be representatives of various self-governmental or administrative bodies (for example provincial or local offices) and of other organizations. The regulations for the establishment of these regional advisory councils were finalized in December 2002.

[6] The NCWM members are nominated by the national organizations of various self-government entities, by academic and other scientific and research entities, and by social, economic, and ecological organizations that relate to water management.

In summary, water policy in Poland, as part of natural resource and environmental policy generally, is based on principles of sustainable and rational resource use. More specifically, the main national targets for integrated water resource management are:

- Improving the quality of surface and underground waters
- Assuring water availability for the population and the economy
- Reducing flood destruction and damage
- Limiting erosion of river banks and bottoms
- Safe operation of hydraulic facilities
- Setting conditions for water use for the power industry, navigation, and recreation.

National policy also requires that these targets be pursued in harmony with social and economic needs as well as with the needs of environmental protection.

Table 10.1 indicates the main organizations and bodies engaged in activities related to water resource management in the Warta basin, with a summary of their main attributions and activities.

10.3
Application of Analytical Framework

10.3.1
Contextual Factors and Initial Conditions

The Warta basin does not feature significant cultural, religious, ethnic, or other divisions within the population that hinder the prospects for successful basin management. Similarly, asymmetries in the distribution of resources among basin stakeholders do not appear to have impeded the move toward the adoption of integrated water resource management at the basin scale.

Economic development of the basin and the country has had effects, however. Poland's agricultural and industrial sectors emerged from the era of Soviet domination lagging behind the West. Support from international financial institutions and from the European Union aided Poland's economic and political transition, and also provided incentives for reforms such as integrated water resource management and the creation of river basin agencies. Still, Poland's economic conditions have led to financial constraints on the government sector, limiting its ability to provide either central funding or revenue autonomy adequate to the tasks of integrated water resource management at the basin scale.

10.3.2
Characteristics of Decentralization Process

The decentralization of government in Poland, and the reform of water policy and water organizations, were attempted over the same (relatively short) period of time, and these simultaneous processes have not always proceeded smoothly. Significant responsibilities for water resource planning and management have been spread across

Table 10.1. Water resource management institutions and roles in the Warta basin

Management level	Institution	Current water management attributions
National	Ministry of Environment, Department of Water Resources	– Under new Water Law, has ceded most responsibility for water management to the NBWM, including water quality standards
	National Board of Water Management (NBWM)	– Established by 2001 Water Law; principal water management entity at national level – Harmonizes RWMA activities and approves their basin plans and progress reports
	National Council of Water Management (NCWM)	– Provides advice to NBWM on such matters as water management, flood control, and drought mitigation
Subnational	Voivodeships (regional administrative entities)	– Various water management responsibilities, including issuing permits for water use and discharge, and their monitoring and enforcement – Fund environmental improvement projects – Manage irrigation facilities and primary drainage systems
	Poviats, starosts (county-level administrative entities)	– In addition to voivodeships, issue permits for water use and discharge (depending on size and potential impact of permit request) in consultation with RWMA – Flood protection and response – Fund environmental improvement projects
	Municipalities	– Water quality monitoring and enforcement (including groundwater) for providers of public water supplies and wastewater services – Fund environmental improvement projects
Basin	Regional Water Management Authority (RWMA) Poznań	– One of seven RWMAs established by 1999 ministerial decree merging DDWMs and RBWMs – On behalf of the central government, performs planning and coordinating functions within basins, oversees the actions of voivodeship and local governments and private users for compatibility with basin water management plans, and maintains specified waterworks and state-owned reservoirs and other facilities
	Regional Council of Water Management (RCWM)	– Established by RWMA; composed of water users and representatives of governmental units in the basin – Approves RWMA basin plan following public consultation

basin and subbasin agencies, and water law reform took several years longer than originally envisioned.

There appear to be no reservations, however, about the central government's commitment to decentralization and democratization reforms, or about its recognition of the local and basin-scale organizations that it created. Central government officials have maintained that commitment throughout the post-Soviet period. However, as noted above, they have held the purse strings rather tightly in light of the limited financial resources available to the public sector in Poland.

10.3.3
Central-Local Relationships and Capacities

Overall, the water law changes in 1997 and 2001, and the merger with the DDWMs in 1999, have given the RWMAs more responsibilities but not additional sources of

revenue. RWMA Poznań had a 2002 budget of US$1.8 million, quite small for an orga-
nization covering such a large basin and employing so many individuals. Of this
allocation, 73.8 percent was used for investments and planning in the basin, 5.9 percent for other development activities, 2.2 percent for water quality activities, 0.1 percent for operations and maintenance, and 18 percent for administration and other
categories. The small amount of financing has left the RWMAs unable to address
the wide array of management concerns within the basin, or even to adequately fund
maintenance of waterworks. The budgetary needs for maintenance and upgrading
of facilities mount each year, and the backlog of needed (but unfunded) tasks accu-
mulates. RWMA Poznań estimates that fulfilling all of its responsibilities would
cost about 100 million zlotys per year; it receives a budget of about 5 million zlotys
per year.

Still, it is important to reiterate how quickly the institutional reforms have tran-
spired. It can be argued strongly that with major reforms occurring regularly since
1990, there has not been enough time for the full implementation of basin management activities, or for a thorough assessment of their performance.

It should also be noted that the water rights system established in Poland
(portions of which predate the democratic transformation) is in certain respects
conducive to integrated water resource management, and the reforms since 1990
have attempted to add a basinwide perspective to that system. Permits for water
use and water discharge are limited in time and quantity, and approved only after
consultation about basin conditions. Fees associated with nonpermitted actions or
with permit violations provide incentives to users and also a revenue source for
environmental improvement projects. Other reforms (such as transferability of
permits) have yet to be undertaken, but most elements of the institutional infrastructure of a water rights system compatible with integrated water resource management
are in place.

10.3.4
Basin-Level Institutional Arrangements

There are basin-level institutions in Poland, but they are only one component of a water
management system that is substantially dispersed, polycentric, and federated.
Although administration, financing, issuing of permits, and other aspects of management are no longer conducted almost entirely at national level, as was the case before
reform, different management functions are distributed, in varying proportions,
among institutions operating at different levels: national, voivodeship, county, municipal,
and basin.

This federal approach, with a sharing of responsibilities across levels and units of
government, allows for the recognition of subbasin communities of interest, and provides overlapping layers of monitoring and enforcement of water management regulations. It does not, however, lend itself to clarity of institutional boundaries or a close
matching of jurisdictional boundaries to basin boundaries. The Warta case provides
a clear reminder that jurisdictional boundary issues will arise within any basin unless all water-related responsibilities are concentrated in the basin management agency,
which is probably infeasible politically if not administratively. This is evident in the
situation of Łódź voivodeship, which is intersected by and divided between two river

basin authorities. It is also evident in the fact that the RWMA in Poznań has to inter-act with several voivodeships that lie partly within and partly outside the Warta basin. These interactions do not yet operate seamlessly, and require all of the actors to understand well their new roles.

Forums for information and communication sharing, and for conflict resolution, are essential in such a polycentric setting. The RCWMs and NCWM appear to be intended to aid in the information sharing and communication roles, but they are so new that there is as yet no record from which to judge their operation. Nor can the effectiveness of conflict resolution methods (which rely strongly on negotiations between governmental units) be assessed conclusively yet.

10.4
Performance Assessment

10.4.1
Stakeholder Involvement

As previously noted, water reform in Poland has mainly been a component of overall reform of government structures following the postwar period of Soviet dominance, and the institutions created in the Warta basin from 1990 to the present – including the RBWM and its successor RWMA Poznań – were created in basins throughout Poland at the same time and in largely identical fashion. The interests and activities of local and regional stakeholders within the basin do not therefore appear to have driven the creation of the principal organizations for management in the Warta basin.

Different constituencies within the basin have different water-related priorities. A number of stakeholders see water quality as a major concern. Most municipalities have switched to groundwater for domestic consumption and have particular interest in maintaining its quality. The regional and national governments' concerns with water quality are associated with the funding of water quality improvement facilities – national and voivodeship funds are the primary source of governmental financial support for improved treatment facilities – and with satisfying national and European Union water quality standards.

Water supply availability and reliability are of principal concern to industrial and agricultural water users in the basin, which rely to a greater degree on surface water supplies than do municipal suppliers. Low flows and drought conditions have the potential to jeopardize the operation of industrial water intakes and irrigation canals.

Flood control is primarily of interest in the urban concentrations along the rivers in the basin, due to the potential injury and economic losses associated with flooding. The national government has taken a particular interest in stimulating better flood protection and response at the regional and local level throughout Poland.

Given this range of interests, RWMA Poznań is logically the organization best placed to manage the difficult task of coordinating potentially conflicting priorities at basin level, and would prefer to have both increased autonomy and the additional funding that would enable it to efficiently discharge its responsibilities. This, however, has not been forthcoming; particularly frustrating is the exclusion of the RWMA from a share of the revenues that flow to the funds for environmental protection and water management at the national, voivodeship, and local levels.

Part of the reason for this combined lack of funding and autonomy lies with opposing subnational and supranational considerations impacting water policy in Poland. The government remains committed to devolution of water resource management to basin level; however, attaining a level of compliance with European Union regulations consistent with ensuring continued access to European Union funding seems best achieved by close centralized supervision and harmonization of RWMA activities (Blaszczyk 2002).

Other tensions can be felt between stakeholders at subnational level. Voivodeship regulators are concerned with maintaining sufficient compliance with national regulations to secure continued funding. Keeping authority over permit issuance, environmental enforcement, and the financing of environmental improvement projects, rather than ceding some control of these functions to RWMAs, are all means to that end. Voivodeship officials can also be presumed to want to avoid the economic losses and the political blame that would accompany a recurrence of a major flood event such as 1997, and thus to support investments in flood control projects.

Provincial and local officials face incentives to grant water use permit applications that will facilitate economic growth, as Poland continues to recover from decades of relative economic stagnation. This creates some difficulty in arresting the growth of groundwater use in urban and industrial areas already experiencing declining groundwater levels. On the other hand, the fee structure (which directs a portion of environmental penalty collections back to the local governments) provides a mild incentive to monitor and enforce environmental compliance.

Agricultural water users want to maintain access to cheap, plentiful water with minimal regulation of use. Keeping the exemption for on-farm water use and water discharge is a logical course of action consistent with that motivation. Industrial water users also can be presumed to want cheap, plentiful water, with minimal regulation and maximum subsidization of waste disposal. Discharge fees are more acceptable to this sector if the revenue supports projects such as treatment plants that help maintain surface water quality and avoid further restrictions on industrial discharges.

Municipal water suppliers have focused on source water quality protection and improvement, which help minimize treatment costs and rate pressures. Of course, rate pressures can also be lessened by subsidization of treatment plants. The latter can be obtained from the funds for environmental protection and water management, primarily at the voivodeship level. Water suppliers operating as enterprise utilities (for example Poznań) also have an interest in seeing tougher regulations on rural and other suppliers, so the full cost recovery rates of urban utilities do not diverge too radically from the costs of water provided by other nearby suppliers.

Environmental interests have seized on the democratization process in Poland since the 1990s to pursue opportunities for representation in a water policy sector that was largely closed to them before. The creation of an NCWM and the RCWMs with a prescribed distribution of seats provides such opportunities. European Union accession, which brings a layer of supranational environmental regulation to bear upon Polish governments, also serves the interests of environmental organizations in Poland.

In summary, recent developments related to water resource management in Poland present opportunities for increased stakeholder involvement at all levels, but it remains unclear whether the sometimes conflicting motives of stakeholders can be coordinated into an efficient and integrated management system.

10.4.2
Developing Institutions for Integrated Water Resource Management

As described in the previous section, despite the nationwide system of RWMAs water management authority and responsibility is far from being integrated at the basin scale. The RWMAs have numerous tasks and responsibilities with respect to water planning and management, facilities operation and maintenance, and coordination and consultation with respect to water use, water quality, and flood control, but decisionmaking authority and funding for several of those tasks have been assigned elsewhere in the Polish governmental system.

An emerging federal system in Poland distributes authority to a number of governmental levels and types (Tonderski and Blomquist 2003). Democratic Poland in the 1990s reorganized and strengthened its provincial and local governments, decentralizing a number of governmental services and functions, establishing fewer but more powerful provincial governments, and resurrecting municipal governments that had been all but destroyed during the Soviet era. That process, which occurred over the same period as the establishment of river basin authorities and the reform of Polish water law and policy, resulted in the spreading of authority for several aspects of water resource management across several levels of government.

Institutional development in the water sector is closely related to the economic instruments available for water management. Fees and penalties are collected from water users and wastewater dischargers. Those revenues are not distributed to or retained by the RWMAs, but are apportioned among the national, provincial (voivodeship), county (poviat or starost), and municipal levels of government. The revenues are available for projects to improve environmental conditions, including water resource conditions, and are disbursed through a system of funds for environmental protection and water management. The funds have their own governance and administrative structure and personnel; RWMAs can provide advice about water improvement priorities within their respective basins, but the choice of projects to assist lies with the funds. A substantial portion of the revenues received by the funds has been devoted to water quality improvement projects, especially treatment plants, but there is no institutional arrangement for prioritizing and funding projects at the basin scale.[7]

---

[7] Additional sources of funding for environmental improvement projects include the Bank for Environmental Protection, which offers preferential credit for environmental protection and water management projects, in cooperation with the national and provincial funds for environmental protection and water management; the Rural Areas Aid Foundation; the Polish Agency for Regional Development; the European Fund for the Development of Polish Rural Areas; and the Foundation of Aid Programs for Agriculture. Poland also receives financial aid for water projects under bilateral agreements and through international organizations (for example the International Monetary Fund, the World Bank Group, and the European Union).

Water quality standards (which are being revised in conjunction with European Union accession) have been established by the Ministry of Environment, and under the new Water Law will be shifted to the NBWM and the RWMA directors. Thus the NBWM and RWMAs will have to establish and monitor revised water quality standards, and determine strategies and priorities for improving compliance with them. The other primary element of improving water quality is through the issuance of wastewater discharge permits, the placement of conditions on those permits, and penalties imposed for unauthorized discharges.

The primary means of controlling water demand is through the issuance of water use permits.[8] These permits are issued by county (poviat or starost) or voivodeship officials, depending on the size (volume) of the request and the scope of its potential impacts. Under the new Water Law, RWMAs are required to establish basin plans that include water use priorities and conditions in the basin. The law obliges the counties and voivodeships to follow those priorities and conditions when deciding whether to grant or deny permit applications, and to consult with the RWMA. Monitoring and enforcement of water use permits or unauthorized water uses are the responsibility of municipal and voivodeship officials, not the RWMAs.

Water tariffs are demand management tools that affect water supply availability and reliability. Tariffs are not set by RWMAs but by municipal officials, whether water supply is provided directly by municipal government or by a utility or private company under municipal regulation. In rural areas, irrigation systems and drainage systems are regulated by the voivodeship administration, with funding support from the central government.

Flood control and flood response are municipal and county responsibilities, under the supervision of a voivodeship department of emergency management. The voivodeship office may take over flood response if an incident surpasses municipal and county boundaries. RWMAs provide information on flood hazards and forecasts, maintain and operate reservoir facilities on the rivers, and plan for the construction of additional facilities if needed.

National laws and regulations therefore constitute a framework within which basin and subbasin actors perform their functions, assisted by a range of financial instruments. The system is however complex, and boundaries of responsibility are not always clear. It remains to be seen how well these attributions will function under the provisions of the new Water Law, within the context of accession to the European Union.

10.4.3
Effectiveness and Sustainability

Water reform in the Warta basin is too recent to draw meaningful conclusions about the effectiveness and sustainability of the measures taken thus far, though progress

---

[8] The new Water Law of 2001 continues the provision of the 1974 Water Law, in slightly amended form, exempting individual households and small farms using 5 cubic meters or less per day from needing a water use permit. The exemption avoids the administrative difficulty and cost of bringing so many small users into the permit system.

may be gauged by reference to certain indicators that have been accorded priority in the basin. These include water quality, water supply availability and reliability, and flood control.

Water Quality

Efforts to improve water quality in the Warta basin have met with mixed results, and it remains a great challenge. Direct water quality measures include concentrations of contaminants; indirect water quality measures include indicators such as fish population, size, and species variety. By both kinds of measures, water quality in the Warta basin has improved with respect to some indicators, and worsened in other places or with respect to other indicators.

Waterborne diseases have not been considered a significant issue in Poland for decades. Generally, water supply facilities have provided waters of good hygienic quality. Much effort toward water quality improvement has focused instead on reducing untreated wastewater discharges from urban areas. Recent investments have stimulated the development of additional sewage treatment plant capacity, including the large modern wastewater treatment plant for the Łódź metropolitan area, completed at the end of 2001, which serves 750 000 people. The wastewater treatment plant serving Poznań has been equipped with the same advanced technology.

European Union funds for rural development have been and will be helpful in improving the quality of water discharged in rural areas; industrial and other discharges continue to take their toll on water quality, however, and the stretch of the Warta between the mouths of the Ner and the Prosna is among the river's most polluted portions. Downstream of the Warta inflow, levels of cadmium and nutrients in the sediments of the Oder increase, reflecting agricultural and industrial (especially electronics manufacturing) activities upstream in the Warta (Müller et al. 2002). The high total organic carbon concentrations measured in the Warta, higher than in other major rivers of Europe and North America, should be reduced following completion of the new treatment facilities for Łódź and Poznań.

Fish populations, species diversity, and size and weight provide indirect indicators of water quality improvement or decline. For the most part, data on fish in the Warta River indicate that water quality still needs considerable improvement. In the farthest upper reaches of the Warta River, where water quality is significantly impaired by mining and other industrial discharges, no fish are found (Przybylski 1996). Other sections of the river vary in species richness and diversity, according to local factors; for example, inflow of pollution from the city of Łódź has a significant negative impact on species diversity (Przybylski 1993). Construction of the Jeziorsko dam – which has no fish ladder – has altered the ecology of a large section of the river, contributing to a two- to threefold reduction in the number of fish species in the stretch of river below the dam in the 10 years after the river was impounded. Some effects may be due to a rise in water temperature from the hydroelectric power plant at the dam; also, the water released below the dam has exhibited greater and rising alkalinity, and lower oxygen levels, compared with the water upstream. The greatest effect, however, appears to be the opening and closing of the dam sluices, due to its impact on downstream aquatic and riparian habitat (Penczak et al. 1998; Penczak 1999; Glowacki and Penczak 2000).

Water Supply Availability and Reliability

A top priority in the Warta basin is the development of additional small retention facilities (small and medium-sized reservoirs) for irrigation supplies, flood protection, and electricity. There are areas within the basin that have been in drought condition up to 10 times over the past 30 years, and RWMA Poznań has been targeting those areas for small reservoir improvements. The principal barrier to construction of needed facilities has been lack of funding. A national policy was developed to promote water storage facilities such as small reservoirs, but funding has not been adequate to implement it and the RWMA does not have funds of its own to devote to this purpose.

Installation of water meters, imposition of water tariffs, and updating the water use permit system hold the greatest promise for improving water supply reliability by managing the growth of water demands. RWMA Poznań started promoting water meters in the early 1990s. Meters had such a dramatic downward impact on water consumption that some urban water suppliers complained of the loss of sales. Compliance with the European Union Water Framework Directive will require full-cost pricing, which may reduce consumption even further.

Although tariffs are a useful tool for restraining the growth of water demand, there is one difficulty with the current system of water use permits and water use tariffs in the Warta basin. Tariffs on permit holders are based on actual water use, not on the amount of the permit, so users tend to apply for larger permits than they will really need (since they will be charged only for what they actually use). Water resources in the basin therefore appear to be overappropriated in places where they may not be. Water permits in the Prosna subbasin, for example, already exceed available flow. One suggestion to rationalize this system has been to base fees on permit amounts rather than actual use, but this would reduce the sensitivity of actual use to fee changes.

In urban and industrialized areas of the basin, economic growth has been accompanied by greater reliance on groundwater, contributing to local concerns about groundwater supplies. Declining groundwater levels are now evident in parts of the Warta basin near Poznań. It is not certain whether a sustained overdraft condition exists at present, but applications for new water uses in the Poznań poviat are being reviewed more carefully, and some have been denied.

Further consideration must also be given to the system of water supply restrictions during periods of drought. Under the new Water Law the RWMA director will make decisions in this regard, rather than voivodeship officials, and RWMA Poznań must develop plans for water restrictions during drought. A step in this direction is the development of a dynamic modeling process for assigning priorities among classes of water, tied to their respective vulnerabilities during a drought or other emergencies. This has been developed initially in the Prosna subbasin, because data are available to support this kind of modeling.

Flood Control

Severe flooding occurred in the Warta basin as throughout Poland in 1997, and the risk of flooding has certainly not been eliminated. The central role in the basin's flood protection system is played by Jeziorsko reservoir, completed in 1986 on the border between Łódź and Wielkopolskie voivodeships, which provides good flood protection for Poznań, the basin's largest city.

Within its own catchment, the Noteć does not present significant flooding risks. This is mainly due to the area's land use, which is dominated by low-intensive pastures and meadows. However, below the Noteć's confluence with the Warta, the city of Gorzów, about 200 kilometers downstream of Poznań, is much more exposed to flood risks; its only protection is provided by embankments that are in poor condition. The Prosna River remains unpredictable, with a large flood potential that threatens another large city in the basin, Kalisz. Significant improvement of the situation will be achieved when a planned reservoir is constructed.

10.5
Summary and Conclusions

10.5.1
Review of Basin Management Arrangements

The Warta case illustrates how much institutional creation and policy reform can be accomplished in a relatively short period when a central government makes and sustains a commitment to decentralization and to integrated water resource management. Fifteen years ago, Poland did not have a rational system of water tariffs, wastewater discharge controls, or water resource planning, let alone a set of basin-scale organizations for water management. Now all of these are in place, albeit still quite new, along with bodies at the national, provincial, and local levels for funding water quality improvements and other environmental protection projects.

The Warta case also illustrates, however, the gaps that can emerge between basin management organizations on the one hand and a policy of integrated water resource management on the other. In the period 1989–2001 the central government in Poland attempted to revise and reform the entire structure of general-purpose governments at the provincial and local levels, to decentralize several state functions to those levels, to create and then reorganize its system of basin management agencies, and reform its policy approach to water resource management. While much has been accomplished, institutional boundaries have not always been clear, and some developments have proceeded quite out of phase – principally, the establishment of the basin agencies without a revenue source of their own, without a structure for basin stakeholder representation and participation, and a decade before the passage of the Water Law that largely defines and authorizes their activities.

10.5.2
Future Prospects

Future prospects for the success of water reform, in Poland generally and in the Warta basin in particular, rest to a large extent with the degree to which the current confusion of attributions among national, subnational, and basin-scale organizations can be rationalized, and the amount of authority for water resource management that is ultimately devolved to the RWMAs. Polish water policy has indeed embraced and moved toward integrated water resource management, but the decentralization has spread water management responsibilities and authority across a large number of subbasin entities. Organizational responsibilities and relationships appear to be substantially less integrated than policy. There are requirements for consultation of the RWMAs and conformity to basin plans, but until 2002 (with the creation of the RCWMs) there was no formal structure to integrate the general-purpose governments at the voivodeship and local levels into the RWMAs or vice versa. Currently there is a substantial gap between the basin-scale organizations that have been created in Poland and the activities that comprise integrated water resource management, most of which have been assigned to subbasin governments.

Consider, for example, the contrast between the river basin authorities in Poland and those in Spain, such as the one for the Guadalquivir basin (Chap. 11). The Guadalquivir basin authority has substantially more management responsibility than RWMA Poznań, and is closer to an agency for integrated water resource management. It carries out permitting, monitoring of water use, and monitoring of water quality, functions that in the Polish case continue to be spread among agencies and levels of government.

The Warta case serves as a reminder that integrated water resource management (a policy approach) is one thing, and coordination at the basin scale (an organizational approach) is another. One can be created and not the other, and it is possible for both to be attempted without being matched to one another. Harmonization of water resource management functions thus remains an unfinished agenda item for the Polish water sector as a whole and in the Warta basin specifically.

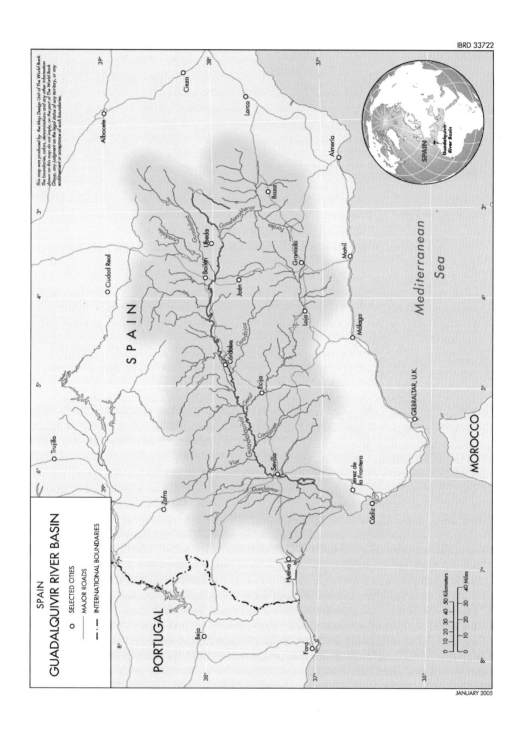

IBRD 33722

SPAIN
GUADALQUIVIR RIVER BASIN

o   SELECTED CITIES
—— MAJOR ROADS
—·—·— INTERNATIONAL BOUNDARIES

0 10 20 30 40 50 Kilometers
0   10   20   30   40 Miles

This map was produced by the Map Design Unit of The World Bank.
The boundaries, colors, denominations and any other information
shown on this map do not imply, on the part of The World Bank
Group, any judgment on the legal status of any territory, or any
endorsement or acceptance of such boundaries.

SPAIN
Guadalquivir
River Basin

PORTUGAL

S P A I N

MOROCCO

Mediterranean Sea

Albacete
Cieza
Lorca
Almería
Baza
Guadiana Menor
Fardes
Úbeda
Granada
Guadiana
Quiebro
Bailén
Motril
Ciudad Real
Jaén
Loja
Málaga
Córdoba
Guadalez
GIBRALTAR, U.K.
Ecija
Genil
Guadalquivir
Corbones
Viar
Sevilla
Jerez de
la Frontera
Trujillo
Cádiz
Zafra
Guadiamar
Huelva
Beja
Faro

JANUARY 2005

Chapter 11

Spain: Guadalquivir Basin

W. Blomquist · C. Giansante · A. Bhat · K. E. Kemper

11.1
Background

11.1.1
Introduction

Spain has perhaps the longest history of any country in developing formal governmental authorities at the river basin scale, with the earliest examples dating from 1926. Although the river basin authorities in Spain (*confederaciones hidrograficas*; CHs) have experienced many changes that have reduced and expanded their responsibilities and their participatory structures over the ensuing 75 or more years, they represent nonetheless a notably long-lived set of basin management institutions.

The basin authority for the Guadalquivir River (Confederacion Hidrografica del Guadalquivir; CH Guadalquivir) is one of the oldest in Spain. The Guadalquivir basin was selected as a case study for this book not only because of CH Guadalquivir's longevity, but because the basin faces the full array of water resource management problems (for example floods as well as drought, tensions between urban and agricultural water uses, and water quality degradation) and is situated almost entirely within the Spanish region of Andalusia (thus minimizing some of the complications of transjurisdictional CHs in Spain).

CH Guadalquivir appears to have been well suited to the tasks that dominated Spanish water policy from the time of its formation to about the 1980s – namely, the construction and operation of water supply and storage facilities with heavy central government subsidies to promote irrigation and hydropower development. As urbanization and industrialization have changed the population and economy of the basin, and as Spanish policy since the 1980s has elevated water demand management, quality protection, and ecosystem health to levels of priority equaling or approximating water supply augmentation, CH Guadalquivir has had difficulty making the transition. Some of the difficulty appears to be inherent in the policy changes themselves, which have added new responsibilities and hence complexities to the management task; some due to organizational inertia; and some due to patterns of stakeholder involvement and stakeholder relationships with basin authority personnel, as exemplified by conflict between the demands of irrigators and those of other concerns.

11.1.2
Basin Characteristics

The Guadalquivir basin extends westerly across southern Spain, and over 90 percent of its area of 57 000 square kilometers lies within the region (communidad autonoma, or CA) of Andalusia. The entire 640-kilometer main stem of the Guadalquivir River itself is located within the CA of Andalusia.

As is the case throughout southern Spain, the Guadalquivir basin has a relatively small share of the nation's water resources, despite having a substantial share of Spain's population. The southern basins of Guadalquivir, Guadiana, Sur, Segura, and Jucar contain 37 percent of Spain's population and represent 41 percent of the Spanish land surface, but receive only 19 percent of the country's total precipitation and runoff.

Water users in the Guadalquivir basin have relied primarily upon surface water resources to supply their needs. There are 52 identified groundwater areas in the basin (Giansante 2003), and groundwater overdrafting is an isolated rather than widespread problem. Precipitation is greatest in the mountains along the edges of the basin and lowest in the valley floor, where most of the population and irrigation are concentrated.

Precipitation and stream flows are highly variable, exposing residents to risks of flooding as well as drought. Mean annual precipitation is about 600 millimeters, but this has ranged from as little as 300 millimeters during drought to as much as 1100 millimeters. Years of high or low precipitation often cluster together, compounding their effects. Seasonal variability is also great, with most precipitation concentrated in the winter months and peak rainfall occurring from November through March. Long, dry summers follow, during which precipitation is virtually nil and evapotranspiration soars.

11.1.3
Water Resource Problems

Water management challenges include flood control and drought protection to sustain the basin's substantial population and significant agricultural production. For the past century, management efforts have focused primarily on regulating river and tributary flows for both flood control and water supply purposes in order to minimize the damage of wet periods and maximize water conservation and storage for dry periods. Current difficulties in water resource management must be assessed against a background of rising population and economic shift giving rise to inevitable conflict over water use.

Irrigation is the dominant consumptive water use in the basin, with 80–85 percent of water consumption. Municipal and industrial uses account for 12–15 percent, with the remaining 5–8 percent apportioned among environmental and other water needs. Each category of water use is increasing, however, and is projected to continue to do so, with little effect on the percentage distribution among uses. Thus, water management authorities do not anticipate that growth of demand in one sector will be met by shifting consumption away from other sectors.

Irrigation areas in the basin, mostly concentrated along the Guadalquivir River stem, were estimated in 1999 to cover 665 000 hectares and continue to grow (Giansante et al. 2000). Olives, fruit, and vegetables are major categories of agricultural production in the central and eastern portions of the Guadalquivir basin, while in the lower basin (known as Baja Guadalquivir) about 35 000 hectares of rice paddies are cultivated, with an estimated water requirement of about 12 000 cubic meters per hectare.

The basin has approximately 4 million inhabitants and experienced a population growth rate of 5.5 percent between 1986 and 1996, compared to a national growth rate of 3.1 percent. It includes all or part of Andalusia's four most populated provinces – Jaen, Cordoba, Grenada, and Sevilla. The population of Sevilla province – the city of Sevilla and surrounding municipalities – is fast approaching 2 million (CH Guadalquivir 1995).

Population expansion and the growth of urban centers such as Sevilla are connected with a regional economic shift. Service industries, recreation, and tourism have expanded as shares of the region's employment and economic product, especially during the 1990s. Reliance on agriculture as the region's sole defining economic pursuit is decreasing, and water use for the urban economy is rising at a pace equivalent to that of irrigation.

The composition of water uses and the changes therein are linked to the basin's water management challenges:

- Flood risks are of greater consequence now that the basin contains 4 million inhabitants, including nearly 2 million in the downstream province of Sevilla. Industrial and commercial sites along the river further escalate the prospective economic losses from flooding. These changes have intensified the water management challenges of flood prevention, control, and response.
- Drought losses are also worsened by the development of water uses in the basin. The continued expansion of irrigated agriculture strains the basin's water resources during dry periods. Although the transition to higher-value crops grown in orchards and groves has beneficial effects on water use efficiency, it creates an additional economic exposure to drought. Trees and vines represent a substantial capital investment for each farm, which magnifies the potential losses resulting from an extended drought.
- Growing municipal and industrial water use adds another dimension to the drought risk situation. While urban water demands are small compared with irrigation demands, they have a year-round character that is less variable than irrigation demands – certain levels of urban water service must be sustained for public health and sanitation despite temporal reductions in precipitation and stream flows. Also, urban water supplies have higher water quality standards to meet, and droughts negatively affect water quality.

Taking all the above into account, and adding in flow requirements for hydroelectric facilities, the water management challenge in the Guadalquivir basin is substantial and complex. Maintaining adequate river flows to sustain irrigation demands and hydropower requirements in the central and eastern portions of the basin, urban water demands at various locations but especially in and around Sevilla, and agricultural water needs in the lower portion of the basin proves particularly difficult. Even in an average year, water demands in the Guadalquivir basin exceed available supplies. In a drought, conditions are only worse.

Water quality problems in the basin are substantial. As agriculture has expanded in the basin, agricultural runoff (including irrigation tailwater) has contributed to water quality degradation in tributaries and the main river stem. Industrial sites, such as manufacturing plants and food processing facilities, have discharged chemical and other wastes into the river system. The growing urban population generates sewage and wastewater that is also discharged to the river system.

Municipal and industrial pollution sources are now covered by national regulations and European Union directives requiring predischarge treatment, but the construction and operation of treatment facilities has not kept pace with the quantities of waste produced and discharged in the basin. According to the Guadalquivir basin plan (CH Guadalquivir 1995), only 50 percent of municipal wastewater received secondary treatment prior to discharge, and most of the other 50 percent received no treatment at all.

11.2
Decentralization Process

11.2.1
Pre-reform Arrangements for Water Resource Management

For much of the period since the original Water Law of 1879 the basic unit for management at basin level has been the river basin authority (CH). Details of the historical evolution of CHs, and of the internal organization of CH Guadalquivir, are presented in Giansante 2003. Created by the central government in 1927, CH Guadalquivir was, for much of its existence, a hydrotechnical agency devoted to the construction of dams, reservoirs, and water conveyance facilities, while water law administration and management of water uses were handled by a separate agency. At various times, however, the central government has merged these functions into one basinwide authority (as is currently the case).

11.2.2
Impetus for Reform

The process of reform in water resource management has taken place over decades in Spain generally and in the Guadalquivir basin specifically, though the defining event may be taken as the new Water Law of 1985. The previous national Water Law, having been in place for over a century, was badly outdated, reflecting priorities of a different era and lacking applicability to emerging concerns about water management and water quality. Regulations of new water demands and wastewater discharges had been pasted onto the existing system during the 1960s and 1970s without revision to the basic law. As one set of observers characterized the situation, the scope of regulated water uses rose but the coherence of water policy declined (Costeja et al. 2002:20).

11.2.3
Reform Process

The decentralization of water resource management with the 1985 Water Law must be seen in the context of national-level changes in the structure of government taking place in Spain. The 1978 Constitution established regional governments (CAs) with a wide range of policymaking responsibilities. Those responsibilities include several aspects of natural resource management, environmental and public health protection, and economic development. Those policy areas overlap in several respects with the water management responsibilities of basin authorities such as CH Guadalquivir.

Although the Constitution establishes that central government is responsible for water resource management in river basins shared by more than one region, as is the case with the Guadalquivir basin, the considerable geographic overlap between the boundaries of CA Andalusia and the Guadalquivir basin means that regional policies of the CA effectively cover the Guadalquivir basin. Several offices and departments of the Andalusian regional government have programs or responsibilities relating to water management in the Guadalquivir basin, including the Water Secretariat, the Department of Agriculture and Fisheries, the Department of the Environment, and the Department of Health.

The 1985 Water Law represented a major reform of water policy in Spain. It was not a product of grass-roots reformism, however, having been designed primarily by expert personnel at the national level. Four principles of the 1985 law were (*a*) integrated management of water resources (surface water and groundwater, water quantity, and water quality); (*b*) the river basin as the appropriate unit or scale of management; (*c*) user participation; and (*d*) reliance on water planning to balance social and economic development with ecological sustainability (Alcon Albertos 2002). These principles are intended to guide the fulfillment of the law's purposes, which were:

- Promoting water quality monitoring, protection, and improvement, and providing equitable and reliable mechanisms for funding those improvements
- Bringing groundwater into the system of water use regulation along with surface water, shifting it from private property to public domain status
- Incorporating economic techniques of water management, and implementing greater recovery of water costs from water users
- Strengthening the CHs through integration of responsibilities at the basin level
- Enhancing representation and participation of water users and other stakeholders on the CHs
- Instituting a more coordinated water planning approach, with river basin plans to be reconciled with a national water plan, which in turn would be reconciled with European Union water regulations.

Some of the most important elements of the law were:

- CHs were to draft basin water plans, including a program of action to meet expected water demands in user sectors over 10-year and 20-year planning horizons; and surface water quality objectives for each water body in the basin.
- The definition of waters in the public domain was broadened to include essentially all surface waters and groundwater, with the government, primarily through the CHs, assuming responsibility for regulating use through licensing and registration.
- Priorities among water uses were revised from the 1879 Water Law, and was now ranked as follows: population supply; irrigation; hydropower; other industrial uses; aquaculture; recreation; and navigation.
- Economic tools for the regulation of water were added, including water regulation rates and water discharge taxes for the use of water and the water domain.[1]
- Authority was conferred upon the CHs for the control of discharges to waters, and for the monitoring and management of river water quality.

A set of amendments to the 1985 Water Law was adopted in 1999. These amendments took account of more recent European Union policy and regulations, including the draft Water Framework Directive of 1998, by strengthening public control over water resources, tightening water quality requirements, and recognizing ecological values. The amendments also contained new guidelines for water pricing and new requirements for water metering to facilitate user-based water charges. In addition, holders of water licenses were given new options for water transfers with the intention of improving water use efficiency, especially in water-scarce locations and times, by allowing water to move from lower-valued uses to higher-valued uses in exchange for monetary compensation.

The 1999 amendments did not alter the organizational structure of water management in Spain, but they exemplified the continued movement of Spanish water law and policy toward an integrated water resource management approach, strengthening public control over water use and the protection of water quality and environmental values, while adding economic incentives and some flexibility with regard to water demands and water uses (Costeja et al. 2002:20) The combination represents a continuation of Spanish water policy reform. The revised national water plan presented in 2001, which raised conservation targets, and the announcement of the newly elected government in 2004 that it would cancel a major interbasin water conveyance project, also reflect the reform trend.

11.2.4
Current Situation

A number of organizations and institutions operating at different levels have influence on water resource management within the Guadalquivir basin (Table 11.1). The

---

[1] The water regulation rate is a payment designed to recover investments and cover operation and maintenance costs of state-funded dams and other waterworks. An occupation fee is an annual payment for the occupation of public water domain (for example riverbanks). A discharge fee is an annual payment for the protection and improvement of the quality of rivers to which discharges are made.

central government in Spain has played a very active role in the development and management of water resources since the late 1800s, and remains an influential stakeholder in the Guadalquivir and other river basins. The national government appoints the CH Guadalquivir president, provides about one third of the CH's budget, and has representation on most of the CH boards and councils. In 1996 the main responsibility for water management passed from the Ministry of Public Works, Transport and Environment to the newly created Ministry of the Environment, to which

Table 11.1. Water resource management institutions and roles in the Guadalquivir basin

Management level	Institution	Current water management attributions
Federal government	Ministry of the Environment	Created 1996; took over water management responsibilities from Ministry of Public Works, Transport and Environment
	Directorate General of Hydraulic Works and Water Quality	Within Ministry of Environment; primary responsibility for water management. Prepares regulations for all aspects of water resources; responsible for design and implementation of hydraulic works of general interest; selects nominees to serve as presidents of interregional CHs, such as CH Guadalquivir
	National Water Council	Consultative body on water policy, bringing together representatives of central, regional, and local administrations
	Ministry of Agriculture, Fisheries and Food	Irrigation planning and the development of irrigation improvement schemes
	Ministry of Industry and Energy	Hydroelectricity, mineral and thermal waters
	Ministry of Health and Consumption	Drinking and bathing water quality
Region	CA Andalusia	Funds intrabasin water projects
	Andalusian Water Council	Est. 1994. Forum for public discussion and planning with respect to water; can make policy recommendations
Basin	CH Guadalquivir	Wide range of basin-level responsibilities (ceding authority to national/regional government as appropriate), including construction and operation of infrastructure, planning, monitoring of water resource conditions, water licensing, water transfers
	Irrigation user communities (*comunidades de regantes*)	Legally recognized groups who apportion irrigation water among members. Bylaws approved by CH; represented on CH boards and commissions. Strong governance and enforcement structure ensures their continuity and assists conflict resolution
	Guadalquivir Irrigation Farmers Union (Federación de Regantes del Guadalquivir; Feragua)	Basinwide association of several (generally larger) irrigation user communities, created 1994. Presents unified position to CH Guadalquivir and CA Andalusia on irrigation communities' interests and concerns
	Municipal water service organizations; e.g. EMASESA (Empresa Municipal de Aguas de Sevilla, SA)	Drinking water supply, sanitation, wastewater treatment. Urban water suppliers represented on CH bodies

the interregional CHs relate. Within the ministry, responsibilities for water management lie primarily with the Directorate General of Hydraulic Works and Water Quality.

At the regional level, the most prominent example of CA Andalusia's interest in water resource management in the Guadalquivir basin is its establishment of the Andalusian Water Council in 1994. The council includes a broad range of water stakeholders and continues to meet as a forum for public discussion and planning. It does not have formal decisionmaking authority, but can make policy recommendations. The regional government's support for a forum such as the Andalusian Water Council appears to reflect a desire to develop a leadership role in water policy that the structure of Spanish water management neither explicitly confers nor forbids, and it has in fact funded a number of intrabasin water projects, several related to improvements in subbasin-level water management practices.

The 1985 Water Law recognized 13 CHs, 9 in interregional and 4 in intraregional basins. CHs are not autonomous – they are under the direction of the central government (in the case of interregional basins) or a regional government (in the case of intraregional basins). The principal responsibilities of CH Guadalquivir (and of CHs generally), as established by the 1985 Water Law and 1999 amendments to it, by the Guadalquivir basin plan and national water plan, and by European Union regulations that are implemented at the river basin level, are as follows:

- Construction, financing, and operation of dams and reservoirs for flow regulation and water storage
- Planning, including development of basin plans, collection and analysis of data, and designation of subbasin units
- Monitoring of water resource conditions, including river flow and water quality
- Water licensing, including reviewing and approving applications for water concessions (licenses for use of water in the public domain), which must be consistent with basin plan priorities and objectives
- Approval of transfers of water concessions between authorized and registered water users (a new function, since 1999)
- Enforcement of Spanish and European Union water regulations, unless assigned specifically to another level or unit of government

CH Guadalquivir is organized into staff offices plus a set of boards, councils, or commissions composed of basin stakeholder representatives and CH staff (Table 11.2). The CH president, who is appointed by the Council of Ministers, serves as head of the CH staff and chairs the advisory bodies.

CH administration and operations are funded by a combination of revenues from central government and revenues generated by the CH itself. By law, CHs must cover their own expenditures. CH Guadalquivir reported a 2001 annual budget of US$115 800 000, with 35 percent coming from central government (on the grounds that the CH incurs costs implementing and enforcing national law and regulations), 30 percent from tariffs and taxes on basin water users, and 35 percent from other sources. CH Guadalquivir retains 100 percent of locally generated revenues for use in the basin; none are distributed back to the central government.

Table 11.2. Boards and offices of CH Guadalquivir. *Source:* Giansante et al. 2000

Executive bodies	
President	Appointed by Council of Ministers
Governing board	Headed by CH president. In charge of financial matters, approves action plans, and defines aquifer depletion and groundwater protection areas
Management bodies	
Operation boards	Coordinate management of hydraulic works and water resources in specific catchment areas or hydrogeological units. Composed of representatives of the administration and of water users
Water Users Assembly	Composed of all users that are part of the operation boards. Makes recommendations concerning CH policies for the coordinated management of hydraulic works and water resources throughout the basin
Reservoir Releases Commission	Makes recommendations to president concerning appropriate amounts and timing for release of water from reservoirs
Waterworks commissions	Provide opportunity for water users who will be served by a particular project to receive information and make recommendations
Planning bodies	
Basin Water Council	Approves basin hydrological plan. Composed of representatives of different departments of central and regional governments, technical services, and basin stakeholders
Planning Office	Drafts, monitors, and reviews basin hydrological plan and provides technical support to Basin Water Council

Irrigation water communities first developed over a century ago and now number over a thousand, exerting a strong influence within the basin. They regulate distribution of irrigation water among their members, from whom they collect fees. They are recognized in Spanish law and their bylaws have to be approved by the CHs, to whom they pay a service fee, and on whose boards and commissions they are represented. Irrigation communities are thus both a means of water management and a means of user participation. They have a strong internal structure for enforcing their own regulations and settling disputes, taking considerable weight off the shoulders of CHs and civil courts. A number of the larger irrigation communities have joined the Guadalquivir Irrigation Farmers Union (Feragua), created in 1994. The purpose of this basinwide association is to present a unified position to CH Guadalquivir and CA Andalusia on irrigation communities' interests and concerns.

Arrangements for provision of municipal water services, including drinking water supply, sanitation, and wastewater treatment, are variable. They may be managed directly by, for example, a public utility, or indirectly by contracting out, typically to a private enterprise. An example of the former in the Guadalquivir basin is the large public utility EMASESA (Empresa Municipal de Aguas de Sevilla, SA), which serves the city of Sevilla plus 11 surrounding municipalities. Urban water suppliers pay fees to CH Guadalquivir for its services in delivering supplies, regulating flows, and monitoring water resource conditions. Urban suppliers are represented on several of the CH boards and councils, though they have fewer representatives than the irrigators.

11.3
Application of Analytical Framework

Application of the analytical framework (Chap. 1) to the Guadalquivir case yields the following observations.

11.3.1
Contextual Factors and Initial Conditions

Some contextual factors and initial conditions have affected the emergence and performance of river basin management in the Guadalquivir markedly, others less so. Water management issues and their resolution in the Guadalquivir case do not appear to have been driven by ethnic, religious, or class divisions in Andalusian society. On the other hand, the economic development of the nation and of the region have had notable effects. The very establishment of the CHs, with an emphasis on the construction of waterworks, emanated from national policies to bolster economic development by promoting first the expansion of agriculture and later the expansion of industry. The Guadalquivir basin in particular was poorer and more rural than most of the rest of the country, and these conditions contributed to an emphasis on the expansion and protection of irrigated agriculture as the central element of the region's economic and social life. These contextual factors have shaped the perceptions of many Guadalquivir basin stakeholders and the CH staff about the principal purposes and appropriate focus of basin management.

11.3.2
Characteristics of Decentralization Process

The CHs were created by the central government for its own purposes – neither because of local-level demands for greater autonomy nor because of a central government desire to shed water management responsibilities, but as an organizational device for executing central government policy one river basin at a time. The CHs nonetheless provided a means for stakeholder participation through representation on boards and commissions. The establishment of basin management institutions in Spain thus carried the potential for greater water user involvement, but that was not the principal reason for which they were created. The CHs are best understood as central government agencies with representative components, with the balance between central control and user participation varying over time. Central government officials have established, diminished, and resurrected the user representation components over the life of the CHs.

Having been established and reconstituted from time to time by the central government, these basin-scale institutions enjoy the recognition of central government officials as legitimate water resource management entities. Such recognition has not been accompanied by an extensive devolution of authority, though. The organization, responsibilities, and policy direction of CHs such as that in Guadalquivir are established primarily by direction from Madrid.

It is equally clear from the long history of CHs (over 75 years) that the central government's policy commitments to decentralization, integrated water resource management, and stakeholder participation have not been consistent through central government transitions. As indicated briefly in this chapter and in detail in Giansante 2003, the roles, priorities, and structure of CHs in Spain have varied considerably, from the pre-dictatorship to the dictatorship periods and in the post-dictatorship period.

11.3.3
Central-Local Relationships and Capacities

Water resource management is one of several services or functions in Spain to have gone through decentralization reforms during the 20th and into the 21st century. Devolution of authority to the basin level has undoubtedly been less than complete: entities such as CH Guadalquivir construct basin-level plans, but they must be submitted for national approval and be consistent with the national water plan. CHs collect revenue of their own for some of the services they provide, and do not have to turn it over to the central government, but they also rely on central government funding for functions established and determined by central government officials. CHs have advisory bodies composed of stakeholder representatives, but several of those councils also have central government representatives and the CH president is still a central government designee.

Also, the structures and responsibilities of CHs are set by national laws and decrees, limiting the flexibility and adaptiveness of the institutional arrangements to varying basin conditions and circumstances. In the Guadalquivir case, this lack of flexibility is best reflected in failure to adjust the representation of basin stakeholders on CH boards and councils, for example to reflect the basin's rapid and substantial urbanization over the past three decades. The central government remains free to alter the governance structure or decisionmaking processes of the CHs (and appoint its leadership) with as much or as little stakeholder consultation as it chooses, but the basin stakeholders do not possess a comparable ability to tailor the institutions to their perceptions of needed or appropriate arrangements.

It is possible of course for basin stakeholders to seek change from the national government, a prospect that brings us to the distribution of national-level political power among basin stakeholders. In Spain, irrigators have organizational clout and know how to use it at the national level to protect their rents. The CH waterworks function has been heavily subsidized, creating an artificially cheap production factor for water users. As irrigation is by far the largest water use, irrigators have been the principal beneficiaries of this subsidization. Urban water customers are by far the largest number, industrial water users contribute more to national output, and regional government policymakers have a major stake in the sustainable use of the basin's water resources, yet irrigators have managed to make the old system persist well past the point when the stated national policy shifted from subsidization toward full cost recovery.

11.3.4
Basin-Level Institutional Arrangements

Basin-level governance institutions do exist in this case, and correspond with the geographic boundaries of the river basin. For most of the time from 1927 to the present, however, the basin authority was primarily a waterworks construction and operation agency. CH Guadalquivir did not function as a basin governance entity through much of its existence. Changes in the responsibilities and the structure of the CHs in 1985 and 1987 appear to have been intended to change them into basin governance organizations.

While geographic boundaries are fitted well, institutional boundaries have become unclear. As noted, the Guadalquivir basin overlaps considerably with the boundaries of the Andalusian regional government, and disagreement over allocation of responsibilities has not necessarily been clarified by redefinition of the roles of regional government and of CHs during recent decades.

The potential for conflict between CA policies and CH policies is heightened by social and political factors. The CH's representation and governance structure gives greatest weight to irrigation users, who are the only subwatershed community with formal recognition in both national law and the CH organizational structure, and the larger irrigation communities have strengthened their influence further by speaking and acting collectively through Feragua. The regional government, on the other hand, is elected on a one person, one vote system and therefore its voting base reflects the region's increasingly urban population and economy. To the extent that irrigation and urban water interests concerning basin management clash, those differences may be expressed as divergent views and policies from the CH and the CA.

A number of basin-level forums exist for information sharing and communication but they do not always function effectively. Meetings of the Water Users Assembly have not been frequent; while operations boards and the CH governing board have met more regularly, their representation is not as broad. At the regional level, the Andalusian Water Council appears to have met regularly and has a broad representative structure, but is not coordinated with the river basin authority. At the subbasin level, users' assemblies in the irrigation user communities may draw greater rates of participation but this varies a great deal across the hundreds of communities in the basin.

A related question concerns the availability of conflict resolution mechanisms within the basin. The irrigation communities have irrigation courts to resolve disagreements among water users, but no comparable forum exists at the basin scale. Water users who wish to challenge a decision taken by CH Guadalquivir can go to the constitutional court, but such a challenge can address only whether the CH has exceeded its granted authority, and cannot address such matters as policy direction and resource allocation.

Monitoring of water deliveries, water use, and water quality occurs in the basin, with water quality monitoring performed principally by CH Guadalquivir and by urban water suppliers. Water use by license holders is monitored inconsistently since the water user registry is not complete, and illegal water use is a problem.

11.4
Performance Assessment

11.4.1
Stakeholder Involvement

Effective implementation of integrated water resource management requires communication with and among stakeholders, and their participation in basin management decisions. Although stakeholder representation on the CH boards and councils was expanded in 1987 and 1989, the management structure and internal culture of CH Guadalquivir has been slow to change, with formal decisionmaking authority remaining concentrated in the hands of the CH president and board.

Irrigators in the Guadalquivir basin enjoy a favored position. They receive a disproportionately large share of water, and at a subsidized rate, which poses a problem in periods of water scarcity when cities have insufficient amounts for their populations. This situation has endured partly because formal and informal practices give irrigators a larger share of influence and interaction with CH staff than urban water suppliers and other interests enjoy. A specific example is that on CH operation boards, a large irrigation community (over 60 000 hectares) can have six representatives, but the largest of cities (for example Sevilla, with over a million inhabitants) can have no more than four. Since representation on the operation boards is translated into representation on the governing board, irrigators' interests are systematically weighted more heavily than other interests. This dominance is reinforced by the fact that the CH boards and commissions that have the broadest stakeholder representation, including the Water Users Assembly, meet infrequently, while irrigator representatives such as Feragua maintain close and frequent contact with CH staff and officials.

The 1985 Water Act can be considered an experiment that opened a far-reaching debate, which has included at times a high degree of conflict. Increased conflict is not necessarily an indicator of failure of the system, since open debate (about water or anything else) had been muffled in Spain for decades. The current conflicts concerning water management appear to be taking place outside the CHs, however, rather than finding expression within them. Basin authorities such as CH Guadalquivir may still be perceived as a relatively closed agency serving irrigators' interests, and not yet as the forum within which a broader range of basin stakeholders express their views and determine basin policy direction.

11.4.2
Developing Institutions for Integrated Water Resource Management

A number of factors favor the ongoing existence and evolution of CHs within the Spanish water resource management system, in addition to their venerable historicity. The CHs have acquired considerable international reputation and prestige (after only perhaps the French river basin agencies or Australia's Murray-Darling Commission). On a more pragmatic plane, the presence of the CHs and their evolving responsibilities for integrated water resource management has aided Spain in receiving European Union support for its environmental programs.

There are, however, certain factors that encourage a degree of inertia, slowing institutional development. One such factor has been the central government's interest in maintaining near uniformity of organizational structures among the CHs. Establishing a system of basin authorities does not have to mean specifying every detail of their internal composition (such as the number of seats for each water sector on each CH board and commission), but in Spain it has. Three reasons have contributed to the central government's desire to maintain a high degree of regularity among the CHs. One is administrative simplicity for central government officials who interact with the CHs. Another is the fact that the national government provides substantial funding for the CHs, and officials who fund organizations often wish to exercise some control over them. Third, interbasin water transfers have long been a centerpiece of Spanish water plans, abandoned only very recently. For a central government that plans (and would fund) movement of water from one basin to another, there may be an understandable desire to harmonize the structure and functions of the basin authorities that would be on the donor and recipient ends of such transfers.

CH officials and staff might prefer to avoid heavy oversight and direction from Madrid with respect to implementation of water management policy, but some limitation of CH discretion is a worthwhile tradeoff for central government's financial support. Being aligned with Madrid as organizations of the central government also gives CH officials and staff some autonomy from local and regional interests, though the irrigator community remains influential, as the internal organization of CH Guadalquivir provides officials and staff less incentive to work closely with urban and environmental interests.

Irrigators are keen to continue this status quo, maintaining access to water subsidies and keeping irrigation's high priority in Spanish policy governing water allocation. They can also be expected to resist mandates to improve efficiency, and have done so successfully thus far (even though some irrigation communities have invested in efficiency improvements). The legal recognition of irrigation communities allows them to police themselves and regulate their own internal conflicts, which is also a valued attribute that irrigators would want to maintain. Keeping a positive working relationship with the CHs can also provide irrigators with a buffer against regional or local government policy initiatives that might be more responsive to the interests and concerns of urban and industrial water users or environmental groups.

Regional and local governments, on the other hand, would prefer to see CHs more aggressively carry out their newer roles in water quality protection and demand management. At one time, the CHs might have been valuable to regional and local governments in water-scarce regions of Spain such as the Guadalquivir basin because the CHs financed and built waterworks and promoted interbasin transfers, but these roles are declining in relevance. River water quality is a greater concern for regional and local governments, both because of its impact on needed (and costly) treatment prior to public consumption and because rivers and coastlines are an attractive amenity for the tourism and recreation industries that have grown in importance in Spain.

11.4.3
Effectiveness and Sustainability

Infrastructure Development

Although their other responsibilities have varied over time, a consistent mission of the CHs has been the construction and operation of waterworks. Even today, 20 years since the waterworks and management functions were recombined in 1985, the waterworks function seems to be what CH Guadalquivir officials and staff are most interested in and comfortable with. It is also the function for which tangible results are most easily measured and achieved. Other management functions – water licensing, demand management – have been performed with less vigor and with less positive results.

The Guadalquivir basin contains over 60 reservoirs. Most were built to provide water for irrigation; some more recently constructed reservoirs were built for urban supply. CH Guadalquivir, like the other Spanish CHs, appears to have succeeded in planning, financing, building, and operating hydraulic works that reduce the water supply variability of the region.

Nevertheless, Guadalquivir's water deficit has not been erased, and exposure to droughts remains a principal problem. Each drought episode of the 20th century was followed by another round of reservoir construction. After the 1981–1983 drought, for example, 18 additional reservoirs were built to provide a 40 percent increase in regulation capacity in the basin. Nevertheless, the 1992–1995 drought left a million Sevilla residents and some irrigation communities in the basin short of water, prompting further reservoir construction, but at high and rising costs.[2]

CH Guadalquivir continues to maintain and promote a long list of additional construction projects within the basin. The effectiveness of the structural approach to addressing supply-demand imbalances is now being called into question for a number of reasons:

- The storage efficiency of individual reservoirs appears to decrease as total reservoir capacity increases. Despite the huge addition to capacity after the drought of the early 1980s, the 1979 record stored water volume (4 000 cubic megameters) has been reached and exceeded only once.
- Siltation is already a significant problem in reservoirs, and eutrophication is a growing problem in reservoirs devoted to urban supply.
- Several reservoirs need significant maintenance or refurbishment to address loss of capacity due to siltation and age. Furthermore, many irrigation canals and distribution systems are aging and in need of maintenance or replacement; likewise the distribution networks of several municipal supply systems.

---

[2] The environmental mitigation measures affiliated with the construction of a planned dam and reservoir site near Sevilla are estimated at 60–70 percent of the total cost of the dam project.

- New water supply availability appears to have simply generated more water demand rather than satisfying the quantity demanded previously. Irrigated land surface has increased (some of it without authorized water use), cities have expanded, industries have grown – all faster than supply augmentation projects have been able to generate additional and reliable water supply.

Flooding also remains an occasional problem, despite the number of facilities that have been built. These facilities have lengthened the interval between damaging floods in the lower basin, but the structural approach has been accompanied by trends that threaten to undermine its success. Urban and agricultural development has encroached into floodplain areas, making the population and economic activities more vulnerable (del Moral and Giansante 2000:100). New efforts to reduce flood damage are focusing on real-time information systems, flood zone designation, and dam safety measures.

Water Quality

Water policy reform added quality protection to the duties of water management agencies in Spain. Water quality objectives are based on intended uses, with quality standards for drinking water sources, for recreational water contact, and for aquatic species.

Municipalities have primary responsibility for urban wastewater treatment. The Andalusian regional government (through the Directorate General of Hydraulic Works) supplies technical and financial assistance to municipalities, and has promoted the establishment of consortia (associations of municipalities) to improve the financial feasibility of building and operating treatment facilities for smaller municipalities. CH Guadalquivir's role has been limited to the collection of a discharge fee from municipalities or businesses, based on the volume and pollution load discharged into rivers. This fee is intended to promote treatment prior to discharge, since the fee diminishes toward zero as effluents are treated. Despite sanctions, however, noncompliance with the discharge fee requirement has been high: in 1992, 30 percent of this tax was left uncollected (MIMAM 1998). The 1999 Water Law amendments transferred fee collection responsibility from the CHs to the Ministry of Finance in the hope of improving performance.

Industrial discharges are mainly concentrated at specific locations, with many related to the food processing industry, notable examples being the effluents derived from the production of olives or olive oil. The olive growing and processing sectors have taken some steps to reduce their discharges to the river system, such as decantation pools or converting to a so-called dual oil pressing system.

Emerging water quality problems in the basin come from nonpoint sources, and thus are closely related to land and water use in the basin. For example, farming on hillsides aggravates water runoff and soil erosion, contributing to the deterioration of stream quality and the siltation of reservoirs. Agricultural runoff tends to be high in nitrates, which affect water quality in reservoirs and promote eutrophication. Nitrates in drinking water supplies have been linked to a potentially life-threatening illness among newborns. Thus far, no systematic efforts to reduce nonpoint pollution such as that coming from runoff have been implemented in the basin.

Groundwater

Many problems have emerged in implementing the 1985 Water Law's integration of groundwater pumping into the water licensing system. In the Guadalquivir basin, approximately 20 000 wells out of a total of about 100 000 remain to be settled, representing considerable unfinished business 20 years after the law was passed. In order for provisions of the 1999 amendments to work (especially the authorization of water transfers), CHs have to keep registries updated. The Ministry of the Environment has designated the completion and updating of registries as a priority and, if needed, will provide information technology support for the process.

Demand Management and Water Use Efficiency

Urban water suppliers within the basin, especially in the Sevilla metropolitan area, have moved more aggressively to use fees and charges for water demand management and drought response. Higher tariffs have not only limited the growth of water demand in urban areas, but have provided revenue to upgrade inefficient or leaking distribution systems.

Agricultural water use remains comparatively inefficient, and irrigators lack incentives to address the situation. The water yield is 70 percent, due to losses inherent in canal use and overall inefficiency. Drip irrigation systems cut these losses but involve high conversion costs and ongoing operational costs. The persistence of inefficient water use in the irrigation sector is due partly to pricing practices. Most irrigation users pay water tariffs based on land surface area rather than water use, so the tariff does not provide a financial incentive to conserve. Furthermore, the tariff irrigators do pay works out to roughly one twentieth of the amount paid by urban users in the Guadalquivir basin – a significant subsidy of irrigators' use of the water infrastructure. The 1985 Water Law required the payment of storage charges and rates in order to recover investments and cover operation and maintenance costs of state-constructed waterworks, but a high degree of subsidization remains and even the artificially low charges have proved difficult to collect.

11.5
Summary and Conclusions

11.5.1
Review of Basin Management Arrangements

In many ways, Spain is further along than almost any other country in pursuing integrated water resource management with a river basin focus, but its experience also demonstrates that the establishment of basin authorities in and of itself is not a solution. Even under such favorable organizational circumstances, political, economic, legal, and historical factors complicate the development and implementation of basin-scale management. Vast changes in the Spanish political system are intertwined, chronologically and substantively, with the development of current water institutions and policies.

The central government in Spain made an effort, starting 80 years ago, to establish basin-scale structures and use them for water development, water planning, and certain aspects of water management. This was a top-down creation of basin organizations by the central government rather than a bottom-up process initiated by local water users. Even with internal structures providing for stakeholder representation, CHs were established as agencies of the central government and remain so today (at least the interregional ones).

The end of the Franco dictatorship in the 1970s brought about a significant decentralization of the whole political system, including not only democratized national-level politics but also the creation of a new level of regional government, the CAs. The devolution of powers to those regional governments included water-related authorities and responsibilities that further decentralize water management in Spain, but not to the basin scale per se. Thus, it is clear that several aspects of water management in Spain have been decentralized, but this does not coincide in all respects with basin management.

Furthermore, the reform of Spanish water law in the 1980s occurred simultaneously with developments at the supranational scale. The year of the new Water Law, 1985, was also the year when Spain entered the European Community, now the European Union. This has added a new layer of water regulations and other policies (for example regarding agriculture) with important implications for water resources. Whatever else may be said of European Union policies, they can hardly be described as a form of decentralization.

Overall, decentralization in the Spanish water arena has been gradual and complex, with several different components. If one takes the 1978–1985 period as the principal decentralization reform landmark, several changes in that period contributed to the current configuration of river basin management, both in terms of regional devolution and of participation in policymaking:

- The transition to democracy and the approval of the Spanish Constitution in 1978
- The creation of regional governments (CAs) with differing degrees of responsibilities in water management
- The issuance of the 1985 decree reuniting hydraulic and water resource management into the functional responsibilities of the CHs
- The approval of the 1985 Water Act, establishing a modern framework of water management, linking surface and groundwater, encouraging user participation, and conferring a central role to water planning
- The 1987 decree restructuring the internal composition of the CHs and their boards and commissions

Thereafter, the Spanish river basin management regime generally, and the Guadalquivir basin regime in particular, have been called to respond to a broader and often more conflictual policy environment of competing water demands and more stringent water and environmental regulations, at the same time operating in a context of restrained public finances and a growing resistance to traditional forms of water supply development. The 1985 Water Law, 1999 amendments, and 2001 national water plan, combined with the European Union Water Framework Directive, have plainly moved Spanish water policy away from a sole emphasis on supply augmentation and toward the incorporation of additional goals of water quality improvement, water demand management, water use efficiency, and environmental protection. At a systemwide level, Spanish water policy in 2006 much more closely resembles an integrated water resource management approach than it did prior to the mid-1980s. In the Guadalquivir basin, these changes have been accompanied by greater friction between urban and irrigation water constituencies (particularly during and after the 1992–1995 drought), and the growing interest of the Andalusian regional government in providing an alternative forum for water policymaking.

The CHs remain central figures of river basin management in Spain, and water policy reform has resulted mainly in the addition of new and different responsibilities for them rather than a transformation of the organizational approach to basin management. As the water management portfolio of the CHs has broadened, they have not performed the new functions as well as their traditional ones, and this contributed to some of the performance outcomes. There has been some involvement of regional and local governments in water policy and politics, and in the Guadalquivir basin even the creation of an alternative structure such as the Andalusian Water Council. The national government has nonetheless remained committed thus far to the CHs, and the CHs are understandably reluctant to cede control over any part of the public water domain now that they have it.

Involvement of other stakeholders and the shift to integrated water resource management have fragmented the configuration of water management organizations and their roles. Costeja et al. (2002:18–21) have described the fragmentation as both horizontal and vertical. The horizontal fragmentation has to do with the dispersion of water policy across more agencies with water-related responsibilities. More specifically, they contend that new aspects of water policy (for example environmental protection, water quality) have spurred the development of new agencies and the assignment of new responsibilities to others. At the national level, in addition to the CHs there are at least seven entities (including ministries and councils) with significant water-related responsibilities. The water-related responsibilities of the Andalusian regional government are distributed among at least five entities (including departments and councils).

The regional governments are also an element of what Costeja et al. (2002) have called the vertical fragmentation of Spanish water resource policy and management. In the pre-reform era, for better or worse, the national government was the dominant water resource policymaker. In the post-reform era, water policy has become a matter of subnational (CA) and supranational (European Union) involvement as well, and the national government is no longer the sole director. The array of subjects that are included in water policy considerations has changed in the direction of integrated water resource management, but the formation of water policy and the conduct of water management are spread across a larger number of entities at different governmental levels.

These changes have certainly challenged the dominance of the traditional water policy community in Spain and helped overcome the exclusion of other constituencies and considerations from meaningful participation. In that respect, it is possible to say that basin management has moved in directions of both decentralization and integration. But the transformation of the Spanish political system, the addition of European Union policies and regulations, and the dispersion of responsibilities into overlapping agencies and levels of government have made for an uncertain transition and, at least in the Guadalquivir basin, generated a mixed record of institutional performance thus far.

11.5.2
Future Prospects

The Guadalquivir case is plainly not one where the central government has recognized basin-level organizations and then abandoned any further commitment to or responsibility for water management. If anything, central government policy reforms since the 1970s have broadened the portfolio of water resource management responsibilities of CHs in ways that CH Guadalquivir staff have found difficult to sustain.

The central government's new water rights regime (begun in 1985 and modified somewhat in 1999) is certainly beneficial from an integrated water resource management standpoint, bringing more users into the system (through the expanded definition of "public domain"), quantifying licenses, maintaining a registry of users, and creating opportunities for trading. The new regime's actual effects on integrated water resource management in the Guadalquivir basin or any other Spanish river basin will depend on its implementation and enforcement, which have been delegated to the CHs.

A final question concerns whether the national government's reforms and the basin institutions have had sufficient time to implement and adapt basin management activities and assess performance. At least with respect to the 1985 Water Law, the answer would seem to be yes, so failure to have an adequate registry of water licenses or control of illegal water uses cannot be attributed to lack of time. On the other hand, the 1999 reforms (encouraging demand management and allowing water transfers) may not have had a long enough period of implementation for accurate assessment.

Part III
Conclusion

Chapter 12

River Basin Management: Conclusions and Implications

W. Blomquist · A. Dinar · K. E. Kemper

What has been learned from this worldwide survey of river basin organizations and these eight case studies? What can be said, based on this study, about integrated river basin management at the lowest possible level – when and why does it work or not work?

This study was stimulated by two recommendations in the water resource management literature. The first recommendation is the *integration* of water resource management at the river basin level, which has been advocated in order to achieve improved management through a more comprehensive approach that combines supply allocation, demand management, quality protection, and ecological preservation or restoration. The second recommendation is the *decentralization* of river basin management to the lowest level, which has been advocated in order to achieve improved management through better adaptation to local conditions, better use of local knowledge and institutions, and greater involvement of stakeholders from the range of sectors involved in water use.

In many locations around the world, pursuing both goals – integration and decentralization – has led to the establishment of river basin organizations. These organizations became the focus of this study. Examining river basin organizations is one way to pose and answer questions such as:

- Is integrated water resource management being pursued at the river basin scale, and with what results?
- Is stakeholder participation occurring at the river basin scale, and with what results?
- Why does river basin management work or not – what factors appear to promote successful implementation of integration and decentralization in river basins, and what factors inhibit it?

As described in Chap. 1, these questions were posed through a combination of a survey of river basin organizations worldwide and a comparative study of eight river basin cases, and the results of that work have been presented in Chaps. 2–11. This final chapter summarizes the overall findings and conclusions from the study, and discusses their implications for water resource policy and for further research.

The conclusions and implications can be grouped into four main categories: (*a*) the duration of the processes of developing river basin management programs and institutions; (*b*) the importance of contextual factors and the need to adapt basin management arrangements to local conditions; (*c*) the wide range of settings in which successful river basin management is possible; and (*d*) the benefits of a partnership approach involving central government resources and offices, regional and local government entities, and water users. These themes are discussed in the four sections that follow; each discussion is followed by summary statements of conclusions, policy implications, and research implications.

12.1
Patience Is a Virtue

The process of institutional development in the river basins covered by this study has been gradual and sometimes difficult, even in basins with considerable reported success in achieving improved water resource management and greater stakeholder involvement. Despite the apparent consensus of water resource professionals about the desirability of integrated water resource management, stakeholder participation, and river basin management, the combination of integration and decentralization is unlikely to be achieved quickly. Cases such as the Guadalquivir and Murray-Darling basins exhibit institutional adaptation spanning nearly a century. Most of the river basin organizations initiated more recently – for example, in the 1980s and 1990s – are still evolving toward an integrated water resource management approach, or building up stakeholder involvement, or both. Longevity by itself is no guarantee of success, but there does appear to be a correlation between the duration of river basin institutions and their development of stakeholder participation and implementation of an integrated approach. In the survey of basin organizations, basins where management reforms had taken longer were also more likely to be perceived as successful and had higher reported levels of stakeholder compliance with water tariffs and local support of basin organization budgets, although lengthy processes that were accompanied by high levels of political conflict negatively impacted perceptions of success.

Science and technology have certainly accelerated humans' ability to learn about the characteristics of physical systems – satellite imagery, electronic databases, computerized basin models, and the like make it possible to map and study river basins much more rapidly. However, the processes of trust building among water users and among public officials, of negotiating organizational structures and authorities, and of working out shared responsibilities for funding and roles in decisionmaking can be accelerated or truncated only at the risk of later institutional failure. For these reasons, the emergence of successful basin management institutions is likely to be a slow process, and patience will be needed from central officials as well as local inhabitants. Reforms that are attempted without sustained commitment to government involvement over a long time frame may not be very effective.

An additional element that follows from the length of the river basin organization development process is the need to institutionalize basin management initiatives rather than relying too much upon one or more charismatic individuals ("champions"). Individual leaders are often critical to getting a river basin management effort started, and attracting resources to support the effort. The importance of champions seems to merit further attention. If building effective river basin management can take decades, however, reliance upon a charismatic champion is risky. While champions did play an important role in a number of the case study basins and certainly had a major impact on the decentralization process, it could also be seen that they do not last forever. Accordingly, early institutionalization of the new institutional setup is crucial, so that reforms can be carried out even when key actors change. Mechanisms for providing sustained funding, securing the participation of stakeholder representatives, and assuring communication among affected organizations should be institutionalized as soon as possible so they do not depend excessively on one personality.

Moreover, the institutional aspects of river basin management – the representation and involvement of stakeholders, the design and administration of public and private organizations, the collection and disbursement of revenues – must be allowed to change as new communities of interest are identified, new resource challenges are confronted, and new possibilities arise. In the words of Molden et al. (2000:87), "Institutions are dynamic entities that need to cater to different management demands as water use changes with the progression of time ... A key feature of an effective institutional design is the ability to adapt to changing needs."

Finally, reforming water resource management arrangements necessarily modifies or even disrupts existing institutions and practices. Time is needed for vested interests that are used to the status quo to become acclimated to changes, and to integrate with them. Efforts to speed that process may be ineffective or even counterproductive. Often it will be preferable to move a step or a stage at time.

Results of this study confirm the observations by Ribot (2002:2) that "the decentralization experiment is just beginning. Discourse has rarely been translated into law or practice. Where it has, people need time to understand and invest in it .... Decentralization will require serious effort and time."

Given the long process of institutional development, many research opportunities remain for observing the emergence and modification of river basin organizations and other entities involved with river basin management. Further studies of how agencies that started with a limited water management portfolio evolved toward integrated water resource management, or of how agencies that started with little or no stakeholder involvement developed more extensive and effective means of stakeholder engagement, would be especially useful and interesting complements to the Guadalquivir and Murray-Darling cases presented in this study.

Conclusion. Establishment of successful river basin management often takes decades.

Policy implications. Political culture, vested interests, and established governmental structures and responsibilities all work against rapid change, and progress often has to proceed in steps or stages, so consistent commitment to the creation and implementation of management at the basin scale is vital.

Research implications. As the history of river basin organizations lengthens, much can be learned from further studies of how organizations that were created for limited functions develop a broader water resource management portfolio, and of how those that were created with little stakeholder involvement develop effective means of engaging stakeholders in decisionmaking.

Recognizing that context matters is not the same as arguing that river basins are like snowflakes, each unique. As Ostrom (1990) contended in relation to the management of common-pool resources, it may be possible to identify some "design principles" that successful cases have in common even while recognizing that the institutional details will differ from place to place and from time to time. Principles of institutional design do not have to be one-size-fits-all organizational recipes. Thus, one of the research implications of this study is that researchers may be able to distill and articulate such design principles from further comparative studies.

Conclusion. Adaptation to basin circumstances has been critical to the prospects for success in gaining stakeholder participation and achieving integrated water resource management. Historical development, physical characteristics of the basin, aspects of the political system, and other contextual factors have greatly affected how river basin organizations were formed, how they changed over time, the functions they performed, and so on.

Policy implications. Integrated and participatory management at the river basin level cannot follow a blueprint. The temptation should be resisted to copy and transport a single structure for river basin organizations to varied circumstances.

Research implications. Although the search for a single best organizational structure is unlikely to be fruitful, further research may yield additional insight into more general design principles of river basin management. Such a program of research can benefit from additional comparative studies of more successful and less successful efforts at basin management around the world.

12.3
Possibilities Are Everywhere

In some countries, there still is a myth that decentralization in river basin management "does not work here because we are different". This position seems to grow out of the belief that there is only one way of managing water resources at the lowest appropriate level, a myth addressed in the previous section. The findings of this study have shown that it is not necessary to copy approaches. The number of institutional options is vast and even small steps in decentralization can lead to improved results in water resource management. Therefore water resource managers, politicians, stakeholders, and international organizations supporting them should look at what works and what fits a certain country or basin and craft institutional arrangements around them.

There is little question that wealthier countries, and wealthier river basins, have some important advantages in creating and sustaining new institutional arrangements for managing water resources at a river basin scale and in securing the involvement of basin stakeholders. It would be a mistake, however, to draw from this obvious generalization the unjustified conclusion that river basin management is not feasible in poor basins and countries. As the data gathered from the survey of river basin organizations show, having wealth or other resource endowments is helpful but not essential to success. There are several examples around the world of basin organizations that have been established, of stakeholder participation occurring, and of water resource conditions improving in impoverished settings.

Of course, success in river basin management may well be slower to develop or face greater obstacles in poor countries or basins. The Jaguaribe and Tárcoles case studies provide examples, especially when viewed in comparison with the more advantageous circumstances of the Fraser or Murray-Darling basins. Nevertheless it remains the case that difficulty is not impossibility; instead, possibilities for establishment of integrated water resource management at the river basin scale with stakeholder involvement are everywhere.

A similar observation holds for the severity of water resource problems in river basins. From the river basin organization survey results and the basin case studies, it does not appear that seriously stressed basins are doomed to failure. Rather, acute water resource crises, or chronic problems of scarcity or flooding, can actually help to stimulate reform activities. In the survey results, the number and severity of water resource problems were actually correlated *positively* with both the initiation of management reforms and the reported success of those reforms. The more ambitious and nearly comprehensive the reform effort was, the more likely users were to report that it was worthwhile and effective.

To be sure, the institutional development process took longer in basins with more severe problems. It also took longer in locations where the water users themselves had to bear more of the financial responsibility for basin management activities. These findings are consistent with the conclusion, however, that circumstances may make basin management more difficult or time consuming in some cases than others without foreclosing the possibility of success. (Indeed, as noted earlier, length of the reform processes actually correlated positively with perceived success, except in cases with high political conflict.)

The implications of these findings for policymakers and for researchers are significant. Policymakers need not be discouraged from supporting basin management efforts even in locations where wealth is limited and water problems are great, provided they understand that progress is likely to be slower and the need for patience greater in those circumstances. Researchers may explore further the threshold conditions of financial capacity and resource degradation that are associated with basin management success, an inquiry that lies beyond the scope of this study.

Conclusion. Successful implementation of river basin management is possible in a wide range of settings, and it does not appear that the poverty of a region or the severity of water resource problems alone has prevented efforts at river basin management.

Policy implications. While wealth and other resource endowments are helpful in establishing new institutions, central government officials and international advisors should not overlook or exclude river basins facing serious water problems or significant poverty. Similarly, local governments and water users should not abandon the pursuit of river basin management because of the poverty of a location or the difficulty of water problems there.

Research implications. This study was not able to explore definitively whether there is some minimal level of financial capacity needed to make river basin management possible, which could be an interesting topic for a differently designed study. Also, the question of whether there are water scarcity, pollution, or other problems so extreme as to be irremediable would have to be addressed by means of some other research design.

12.4
Shared Responsibility Is Essential

A decentralized approach to water resource management does not mean the central government must be removed from the scene, nor does the pursuit of integrated water resource management mean that responsibilities must be organized and fulfilled centrally. Rather, this study supports the view that successful pursuit of river basin management entails a sharing of responsibilities between central government agencies and local communities and water users. Indeed, shared responsibility may prove to be one of those design principles referred to earlier – a common characteristic of successful basin management regimes, though the details of responsibility sharing will differ from one setting to others. Further research could explore and possibly confirm this claim.

The combination of integration and decentralization represented in river basin management calls for a revised distribution of authority and responsibilities among existing public and private organizations, and will usually involve the creation of new organizations as well. Carrying out these rearrangements requires creativity as well as adaptation to the local context. The creativity aspect includes thinking outside of traditional categories: "In many cases, the problem is not so much whether a certain service should be provided by a central, regional, or local government since the service has to be provided with the intervention of all three ... the real challenge is how to organize the joint production of the service" (Prud'homme 1994:31; also Rolla 1998:31). In the river basin management context, this observation may apply to everything from infrastructure maintenance or flood response to water quality monitoring and water supply allocation.

A key element of shared responsibility is the financial support of a river basin organization and of water management functions and services. Budgeting, securing revenues, and deciding upon expenditures are important activities for involvement of basin stakeholders. The basin organization survey yielded an intriguing combination of findings in this respect.

As would be expected, adequate revenues were correlated positively with reported success. Budget and budgetary decisionmaking are of major importance for stakeholders to remain interested in decentralization. Additionally, the autonomy of a river basin organization or other water user organizations to retain revenues generated within the basin for use within the basin was also correlated positively with success. The implication is that the longstanding call in the water resources literature (Mody 2004) and of course in practice by stakeholders, such as for instance in Brazil and in Mexico, to allow that financial resources that have been levied through tariffs in a basin also remain there, is strongly corroborated by this study. Furthermore, and also consistent with scholarly literature on water economics, reported success was higher in basins where stakeholders accepted greater financial responsibility by complying with water tariffs and contributing to the budget for basin management.

At the same time, however, central government financial support was correlated positively with success. Thus it appears that the financial dimensions of river basin management are both important and complex: success is associated with central government support, with water user financial responsibility, and with autonomy to retain and use basin-generated revenues within the basin.

The basin case studies reinforced the findings from the river basin organization survey on these matters. As described in Chap. 2 as well as in the individual case study chapters, adequacy and consistency of central government financial support for water resource management activities and for basin organizations themselves had a great deal to do with which cases had progressed relatively smoothly and which ones had experienced difficulties. Yet the financial participation of stakeholders was just as important. What appeared to work best was a mix of central government and basin-generated funding – decentralizing water resource management to the river basin level does not have to mean that local users are on their own, nor does it require the central government to foot the entire bill.

One of the most interesting implications of the study findings is that governments have little to lose with decentralization. There are some cases where governments are backtracking on reforms that had already been implemented or where reforms simply do not get past the political hurdles in the first place. However, the findings clearly show that determined, strong government support, including financial support, will remain important even in the very long run. This implies that while governments do give up some decisionmaking power, they will effectively always have to be a partner in water resource management. Resistance to devolution of power could therefore be much less than in many places currently is the case.

The lessons learned with respect to funding appear to be part of a broader lesson about the role of central government in institutional reform. "Decentralizing, while promoting the role of local government, does not render the central state illegitimate, but requires a different qualification of its role" (Rolla 1998:33; also Andersson and Gibson 2005:22). There is an important distinction between central government support or involvement on the one hand, and central government direction on the other. A central government's policy of decentralization has to be more than mere rhetoric in order for stakeholder involvement to be meaningful and for the promises of integrated water resource management to be realized. Real empowerment and leadership by local communities, and the implementation of subsidiarity principles, are vital (Ribot 2002; Rolla 1998). In the basin organization survey, while central government support was positively related to reported success, top-down direction of the decentralization process was negatively related to reported success. The combination of roles between central agencies or officials and local communities needs to be worked out in a manner that falls somewhere between central government abandonment and central government control.

Such a partnership arrangement is certain to require adaptation and learning over time. Researchers exploring other combinations of cases in the future will undoubtedly learn more about how the distribution of roles and functions involved in river basin management evolve. In this study, longer-lived cases such as the Guadalquivir and Murray-Darling basins, and even more recent ones such as the Alto Tietê and

Fraser basins, exhibited a number of instances when basin management participants deliberately realigned the mix of roles and functions performed by central government agencies, local communities, and the basin organizations themselves.

The ongoing practice of institutional entrepreneurship – crafting new arrangements and modifying existing ones – is more than a necessary response to new knowledge and changed conditions. It is itself a form of learning and skill development needed by human beings as they interact with each other and the environment. This is an important conclusion in its own right.

Conclusion. Successful implementation of river basin management was more common in settings where water users, local and regional governments, and central government officials shared responsibility for planning, funding, and executing basin management functions.

Policy implications. River basin management can work as a partnership of central government with local or regional communities of interest. It should not be regarded as a way for central government to abandon support or responsibility in the area of water resource management, or for local interests to gain or perpetuate subsidization of water management infrastructure and services.

Research implications. This conclusion would bear examination across time as well as across cases. Researchers could examine how the distribution of responsibilities among local and national, as well as private and public, entities evolves over the life of a river basin management effort.

There remains plenty to be learned about river basin management, decentralization, integration, and institutions. This study has contributed a framework for analyzing these phenomena, a statistical analysis of survey data from a worldwide sample of river basin organizations, and a comparative case study of a small number of cases that vary across a number of dimensions of initiation, duration, structure, and performance. Despite these contributions, and perhaps to some degree because of them, the agenda for policymakers and researchers remains full.

References

Abers RN, Keck ME (2005) Águas Turbulentas: Instituições e Práticas Políticas na Reforma do Sistema de Gestão da Água no Brasil. In: Melo MA, Lubambo CW, Coelho CB (eds) Desenho Institucional e Participação Política: Experiências no Brasil Contemporâneo. Editora Vozes, Rio de Janeiro, Brazil

Agência da Bacia do Alto Tietê (2004) Quadro da Situação da Bacia do Alto Tietê em Novembro de 2004. São Paulo, Brazil (available at *www.agenciaaltotiete.org.br*)

Agrawal A (2002) Decentralization policies and the government of the environment. Polycentric Circles 8(1):4–5

Alaerts GJ (1999) Institutions for river basin management: The role of external support agencies (international donors) in developing cooperative arrangements. Paper presented at International Workshop on River Basin Management – Best Management Practices, Delft University of Technology/River Basin Administration (RBA), the Hague, Netherlands, 27–29 October 1999

Alcon Albertos C (2002) The Spanish Confederaciones Hidrograficas: Historical development and the future scope. Presentation at the International Conference of Basin Organizations, Madrid, Spain, November 2002

Alvim ATB (2003) A Contribuição do Comitê do Alto Tietê à Gestão da Bacia Metropolitana, 1994–2001. PhD in Urban and Environmental Structures, Faculdade de Arquitetura e Urbanismo, Universidade de São Paulo, Brazil

ANA (Agência Nacional de Águas) (2002) Overview of hydrographic regions in Brazil. ANA, Brasília, Brazil

Andersson KP, Clark CG (2005) To have a say and a saw: Is decentralization good for the forest? Paper presented at a conference in honor of Elinor Ostrom, Bloomington, Indiana, 22–23 November 2005

Azevedo, LGT, Asad M (2000) The political process behind the implementation of bulk water pricing in Brazil. In: Dinar A (ed) The political economy of water pricing reforms. World Bank/Oxford University Press, Washington, DC, USA, pp 339–358

Ballestero A (2004) Institutional adaptation and water reform in Ceará. Master's thesis, School of Natural Resources and Environment, University of Michigan, Ann Arbor, Michigan, USA

Ballestero M (2003) Tárcoles basin. World Bank, Washington, DC, USA (Case Study Background Paper)

Baltar AM, Azevedo LGT, Rêgo M, Porto RLL (2003) Sistemas de Suporte à Decisão para a Outorga de Direitos de Uso da Água no Brasil. World Bank, Brasília, Brazil (Série Água Brasil 2)

Blaszczyk P (2002) The challenge of implementing the Water Framework Directive in Poland. United Nations Economic Commission for Europe, Geneva, Switzerland (available at *www.unece.org/env/ water/documents/Poland.pdf*)

Blomquist W (1988) Getting out of the commons trap: Variables, process, and results in four groundwater basins. Social Science Perspectives Journal 2(4):16–44

Blomquist W, Dinar A, Kemper KE (2005) Comparison of institutional arrangements for river basin management in eight basins. World Bank, Washington, DC, USA (Draft Working Paper)

Bromley DW (1989) Economic interests and institutions. Basil Blackwell, New York, USA

Calbick KS, McAllister R, Marshall D, Litke S (2004) The Fraser River basin, British Columbia, Canada. World Bank, Washington, DC, USA (Case Study Background Paper, available at *lnweb18.worldbank.org/ ESSD/ardext.nsf/18ByDocName/SectorsandThemesRiverBasinManagementIntegratedRiverBasin ManagementProject*)

Carmigmani A (2000) A Água como Recursos Natural e os Usos a Serem Onerados: A Posição da SABESP sobre o PL 20/98. In: de M. Thame AC (ed) A Cobrança pelo Uso da Água. IQUAL Editora, São Paulo, Brazil

Castro JE (1995) Decentralization and modernization in Mexico: The management of water services. Nat Resour J 35:461–487

Cerniglia F (2003) Decentralization in the public sector: Quantitative aspects in federal and unitary countries. J Policy Model 25:749–776

CH Guadalquivir (1995) Memoria del Plan Hidrológico del Guadalquivir. Confederación Hidrográfica del Guadalquivir, Sevilla, Spain

COGERH (Companhia de Gestão dos Recursos Hídricos) (2000) VII Seminário de Operação dos Vales do Jaguaribe e Banabuiú. COGERH, Secretaria dos Recursos Hídricos, Governo do Ceará, Fortaleza, Ceará, Brazil (unpublished document, *www.cogerh.com.br*)

COGERH (Companhia de Gestão dos Recursos Hídricos) (2001) Relatório do VIII Seminário de Planejamento e Operação Vales do Jaguaribe e Banabuiú. COGERH, Secretaria dos Recursos Hídricos, Governo do Ceará, Fortaleza, Ceará, Brazil (unpublished document, *www.cogerh.com.br*)

COGERH (Companhia de Gestão dos Recursos Hídricos) (2002) Relatório do XIX Seminário de Operação dos Vales do Jaguaribe e Banabuiú. COGERH, Secretaria dos Recursos Hídricos, Governo do Ceará, Fortaleza, Ceará, Brazil (unpublished document, *www.cogerh.com.br*)

COGERH (Companhia de Gestão dos Recursos Hídricos) (n.d.) Águas do Vale: Plano de Uso Racional da Águas nos Vales do Jaguaribe and Banabuiú. COGERH, Secretaria dos Recursos Hídricos, Governo do Ceará, Fortaleza, Ceará, Brazil (unpublished document)

COGERH (Companhia de Gestão dos Recursos Hídricos)/Engesoft (1999a) Plano de Gerenciamento das Águas da Bacia do Rio Jaguaribe. Fase 1 (Diagnóstico). Fortaleza, Ceará, Brazil (Vol 1: Estudos de Base de Hidrologia, Tomo I: Atualização e Análise de Dados Hidrometeorológicos)

COGERH (Companhia de Gestão dos Recursos Hídricos)/Engesoft (1999b) Plano de Gerenciamento das Águas da Bacia do Rio Jaguaribe. Fase 1 (Diagnóstico). Fortaleza, Ceará, Brazil (Vol 1: Estudos de Base de Hidrologia, Tomo II: Estudos de Oferta Hídrica)

COGERH (Companhia de Gestão dos Recursos Hídricos)/Engesoft (1999c) Plano de Gerenciamento das Águas da Bacia do Rio Jaguaribe. Fase 1 (Diagnóstico). Fortaleza, Ceará, Brazil (Vol 1: Estudos de Base de Hidrologia, Tomo III: Estudo do Impacto Cumulativo da Pequena Açudagem)

COGERH (Companhia de Gestão dos Recursos Hídricos)/Engesoft (1999d) Plano de Gerenciamento das Águas da Bacia do Rio Jaguaribe. Fase 1 (Diagnóstico). Fortaleza, Ceará, Brazil (Vol 2: Estudos de Demanda)

COGERH (Companhia de Gestão dos Recursos Hídricos)/Engesoft (1999e) Plano de Gerenciamento das Águas da Bacia do Rio Jaguaribe. Fase 1 (Diagnóstico). Fortaleza, Ceará, Brazil (Vol 3: Estudos de Balanço Oferta X Demanda)

COGERH (Companhia de Gestão dos Recursos Hídricos)/Engesoft (1999f) Plano de Gerenciamento das Águas da Bacia do Rio Jaguaribe. Fase 1 (Diagnóstico). Fortaleza, Ceará, Brazil (Vol 4: Estudos Ambientais)

COGERH (Companhia de Gestão dos Recursos Hídricos)/Engesoft (1999g) Plano de Gerenciamento das Águas da Bacia do Rio Jaguaribe. Fase 3 (Programas de Ações). Fortaleza, Ceará, Brazil (Vol 2: Programa de Abastecimento dos Núcleos Urbanos)

CORHI (Comitê Coordenador do Plano Estadual de Recursos Hídricos) (1997) Simulação da Cobrança pelo Uso da Água: Relatório Preliminar do Grupo de Trabalho-Modelo de Simulação para o CRH. São Paulo, Brazil

Costeja M, Font N, Rigol A, Subirats J (2002) The evolution of the national water regime in Spain. EUwareness Report. Universitat Autonoma de Barcelona, Barcelona, Spain

Curtis A, Shindler B, Wright A (2002) Sustaining local watershed initiatives: Lessons from landcare and watershed councils. J Am Water Resour As 38(5):1207–1216

da Cunha FM (2004) Desempenho Institucional na Gestão de Recursos Hídricos: O Caso dos Subcomitês de Bacia Hidrográfica Cotia-Guarapiranga e Billings-Tamanduateí na Região Metropolitana de São Paulo. Msc. in Environmental Science, PROCAM/Universidade de São Paulo, Brazil

da Silva FOE (2003) Gestão de Recursos Hídricos: Outorga e Cobrança pelo Uso da Água. A Experiência do Estado do Ceará. PowerPoint presentation at the Agência Nacional de Águas, Brazil

del Moral L, Giansante C (2000) Constraints to drought contingency planning in Spain: The hydraulic paradigm and the case of Seville. Journal of Contingencies and Crisis Management 8(2):93–102

Dinar S, DinarA (2005) Scarperation: The role of scarcity in cooperation among basin riparians. Paper presented at the 2005 International Study Association Convention, Honolulu, Hawaii, 1–5 March 2005

Dinar A, Kemper KE, Blomquist W, Diez M, Sine G, Fru W (2005) Decentralization of river basin management: A global analysis. World Bank, Washington, DC, USA (Draft Working Paper)

Dorcey A (1990) Sustainable development of the Fraser River estuary: Success amidst failure. Paper prepared for the Coastal Resources Management Group, Environment Directorate, OECD, Paris, France

Dulude J (2000) Tackling the Tárcoles: Tough task. The Tico Times Online 7 July 2000 (*www.ticotimes.net/archive/07_07_00-2.htm*)

Easter KW, Hearne RR (1993) Decentralizing water resource management: Economic incentives, account-ability, and assurance. World Bank, Washington, DC, USA (Policy Research Working Paper 1219)

Formiga Johnsson RM (1998) Les eaux brésiliennes: Analyse du passage à une gestion integrée dans l'état de São Paulo. PhD in Environmental Sciences and Techniques, Université de Paris XII, Val de Marne, France

Formiga Johnsson RM (2004) Brazilian case studies: Jaguaribe and Alto Tietê River basins. World Bank, Washington, DC, USA (Case Study Background Paper)

Formiga Johnsson RM, Lopes PD (eds) (2003) Projeto Marca d'Água: Seguindo as Mudanças na Gestão das Bacias Hidrográficas do Brasil. Caderno 1: Retratos 3x4 das Bacias Pesquisadas. FINATEC/Universidade de Brasília-UnB, Brasília, Brazil

Formiga Johnsson RM, Campos JD, Canedo de Magalhães P et al. (2003) A Construção do Pacto em Torno da Cobrança pelo Uso da Água na Bacia do Rio Paraíba do Sul. Proceedings of XV Simpósio Brasileiro de Recursos Hídricos: Desafios à Gestão da Água no Limiar do Século XXI, Curitiba (PR), Brazil, 23-27 November 2003

Fraser Basin Council (1997) Charter for sustainability. Fraser Basin Council, Vancouver, British Columbia, Canada

FREMP (Fraser River Estuary Management Program) (1996) The Fraser River estuary: Environmental quality report. FREMP, Burnaby, British Columbia, Canada

FUSP (Fundação Universidade de São Paulo) (2001) Plano da Bacia do Alto Tietê: Relatório Final, Versão 1.0. Comitê da Bacia Hidrográfica do Alto Tietê, São Paulo, Brazil

FUSP (Fundação Universidade de São Paulo) (2002a) Plano de Bacia do Alto Tietê: Relatório Final, Versão 2.0. Comitê da Bacia Hidrográfica do Alto Tietê, São Paulo, Brazil

FUSP (Fundação Universidade de São Paulo) (2002b) Plano de Bacia do Alto Tietê: Sumário Executivo. Comitê da Bacia Hidrográfica do Alto Tietê, São Paulo, Brazil

Galasso E, Ravallion M (2000) Distributional outcomes of a decentralized welfare program. World Bank, Washington, DC, USA (Policy Research Working Paper 2316)

Ganesh KN, Ramakrishnan C (2000) Education and people's planning campaign: The Kerala experience. International Conference on Democratic Decentralization, 23-27 May 2000, State Planning Board, Trivandrum, Kerala, India

Garjulli R (2001a) A Participação dos Usuários na Implementação dos Instrumentos de Gestão de Recursos Hídricos: O Caso do Ceará. IV Diálogo Interamericano de Gerenciamento de Água: Em Busca de Soluções 22-26 April 2001, Foz de Iguaçu, Paraná, Brazil

Garjulli R (2001b) Oficina Temática: Gestão Participativa dos Recursos Hídricos – Relatório Final. PROÁGUA/ANA, Aracaju, Brazil

Garjulli R, de Oliveira JL, da Cunha MAL et al. (2002) Projeto Marca d'Água: Relatório da Bacia do Rio Jaguaribe. Fortaleza, Ceará, Brazil (*www.marcadagua.org.br*)

Giansante C (2003) Guadalquivir basin. World Bank, Washington, DC, USA (Case Study Background Paper, available at *lnweb18.worldbank.org/ESSD/ardext.nsf/18ByDocName/SectorsandThemes RiverBasinManagementIntegratedRiverBasinManagementProject*)

Giansante C, Bakker K, Crook E, van der Grijp NM (2000) Adaptive responses to hydrological risk: An analysis of stakeholders. University of Seville, Seville, Spain (SIRCH Working Paper 6)

Glowacki L, Penczak T (2000) Impoundment impact on fish in the Warta river: Species richness and sample size in the rarefaction method. J Fish Biol 57(1):99-108

Gomes L de C (2004) Outorga de Recursos Hídricos no Estado de São Paulo. Departamento de Água e Energia Elétrica (DAEE), São Paulo, Brazil (PowerPoint presentation, available at *www.daee.sp.gov.br*)

Goss K (2003) Comprehensive water management in the Murray-Darling basin. Proceedings of River Engineering, JSCE 8:1-6

Grey D, Sadoff CW (2006) Water for growth and development. In: Comision Nacional del Agua (ed) Thematic documents of the IV World Water Forum. Comision Nacional del Agua, Mexico City.

GUS (Główny Urzad Statystyczny, Central Statistical Office) (2001) Environment 2001. GUS, Warsaw, Poland

GUS (Główny Urzad Statystyczny, Central Statistical Office) (2002) Statistical yearbook GUS, Warsaw, Poland

Haisman B (2003) Murray-Darling basin, Australia. World Bank, Washington, DC, USA (Case Study Background Paper, available at *lnweb18.worldbank.org/ESSD/ardext.nsf/18ByDocName/ SectorsandThemesRiverBasinManagementIntegratedRiverBasinManagementProject*)

Harrison DC (1981) Basinwide perspective: An approach to the design and analysis of institutions for unified river basin management. In: North RM, Dworsky LB, Allee DJ (eds) Unified river basin management. American Water Resources Association, Minneapolis, USA, pp 427-437

IBGE (Instituto Brasileiro de Geografia e Estatística) (2006) Brasil em Síntese. IBGE, Diretoria de Pesquisa, Departamento de Contas Nacionais, Rio de Janeiro, Brazil (available at *www.ibge.gov.br*)

ICWE (International Conference on Water and the Environment) (1992) The Dublin Statement and report of the conference. International Conference on Water and the Environment, 26–31 January 1992

Keck, ME (2002) Water, water everywhere, nor any drop to drink: Land use and water policy in São Paulo, Brazil. In: Evans P (ed) Livable cities? Urban struggles for livelihood and sustainability. University of California Press, Berkeley, California, pp 161–194

Keck ME, Jacobi P (2002) Projeto Marca d'Água: Relatórios Preliminares 2001. A Bacia Hidrográfica do Alto Tietê, São Paulo, Brazil

Kemper KE (1996) The cost of free water: Water resources allocation and use in the Curu valley, Ceará, Northeast Brazil. Linköping University, Linköping, Sweden (Linköping Studies in Arts and Science 137)

Kemper KE (1998) Institutions for water resource management. In: World Bank (ed) Brazil: Managing pollution problems. World Bank, Washington, DC, USA (The Brown Environmental Agenda, Report No. 16635-BR, vol II: Annexes)

Kemper KE (1999) The water market in the northern Colorado Water Conservancy District – institutional implications. In: Mariño M, Kemper KE (eds) Institutional frameworks in successful water markets: Brazil, Spain, and Colorado, USA. World Bank, Washington, DC, USA (Technical Paper 427)

Kemper KE, Olson DI (2000) Water pricing: The dynamics of institutional change in Mexico and Ceará, Brazil. In: Dinar A (ed) The political economy of water pricing reforms. World Bank/Oxford University Press, Washington, DC, USA, pp 339–358

Kingdon J (1995) Agendas, alternatives, and public policies, 2nd edn. Longman, New York, USA

Kundzewicz ZW, Chalupka M (1994) The need for integrated river basin management: A case study from eastern Europe. In: Kirby C, White WR (eds) Integrated river basin management. John Wiley and Sons, Chichester, UK, pp 483–492

Laboratório de Hidrologia e Estudos do Meio Ambiente/COPPE/Universidade Federal do Rio de Janeiro (2004) Estudo para o Aperfeiçoamento da Metodologia de Cobrança das Bacias dos Rios Paraíba do Sul e Guandu. Fundação COPPETEC, Rio de Janeiro, Brazil (Research Report to FINEP/MCT/CT-HIDRO)

Lemos MC, de Oliveira JLF (2004) Can water reform survive politics? Institutional change and river basin management in Ceará, Northeast Brazil. World Dev 32(12)

Marshall D (1998) Watershed management in British Columbia: The Fraser River basin experience. Environments 25(2/3):64–79

McGreer E, Belzer W (1999) Contaminant sources to the Fraser River basin. In: Gray C, Tuominen T (eds) Health of the Fraser River aquatic ecosystem. Aquatic and Atmospheric Sciences Division, Environment Canada, Vancouver, Canada (Department of Environment Fraser River Action Plan Report 1998-11)

MDBC (Murray-Darling Basin Commission) (1999) Salinity and drainage strategy, ten years on. MDBC, Canberra, Australia

MDBMC (Murray-Darling Basin Ministerial Council) (1987) Salinity and drainage strategy. MDBMC, Canberra, Australia (Background Paper 87/1)

MDBMC (Murray-Darling Basin Ministerial Council) (2002) The Living Murray – discussion paper. MDMBC, Canberra, Australia

MIMAM (Ministry of the Environment) (1998) Libro Blanco del Agua en España, vol I and II. Ministry of the Environment (Ministero de Medio Ambiente, MIMAM), Madrid, Spain

Ministério do Meio Ambiente/Secretaria de Recursos Hídricos (2006) Plano Nacional de Recursos Hídricos. MMA/SRH, Brasília, Brazil (Vol. 1: Panorama e Estado dos Recursos Hídricos do Brasil)

Miranda-Neto A da C, Marcon H (2000) A Água como Recursos Natural e os Usos a Serem Onerados: A Visão da ASSEMAE." In: de M Thame AC (ed) A Cobrança pelo Uso da Água. IQUAL Editora, São Paulo, Brazil

Mody J (2004) Achieving accountability through decentralization: Lessons for integrated river basin management. World Bank, Washington, DC, USA (Policy Research Working Paper 3346)

Molden D, Sakthivadivel R, Samad M (2000) Accounting for changes in water use and the need for institutional adaptation. In: Abernethy CK (ed) Intersectoral management of river basins: Proceedings of an international workshop on integrated water management in water-stressed river basins in developing countries. Loskop Dam, South Africa, 16–21 October 2000, pp 137–158

Müller A, Heininger P, Wessels M, Pelzer J, Grünwald K, Pfitzner S, Berger M (2002) Contaminant levels and ecotoxicological effects in sediments of the River Odra. Acta Hydroch Hydrob 30(5-6): 244–255

Musgrave W (1997) Decentralized mechanisms and institutions for managing water resources: Reflections on experience from Australia. In: Parker D, Tsur Y (eds) Decentralization and coordination of water resource management. Kluwer Academic Publishers, Dordrecht, the Netherlands, pp 429–447

Niemczynowicz J (1992) The Polish environmental syndrome: Environmental problems and ways to approach them. Water Int 17(4):175–186

Norris R, Liston P, Davies N, Coysh J, Dyer F, Linke S, Prosser I, Young B (2001) Snapshot of the Murray-Darling basin river condition. Report to the Murray-Darling Basin Commission, Canberra, Australia

Oechssler J (1997) Decentralization and the coordination problem. J Econ Behav Organ 32:119–135

Oliveira JLF, Garjulli R, da Silva UPA (2001) Conflitos e Astratégias: A Implantação do Comitê de Bacia do Rio do Curu (unpublished document)

Ostrom E (1990) Governing the commons: The evolution of institutions for collective action. Cambridge University Press, New York, USA

Ostrom E (1992) Crafting institutions for self-governing irrigation systems. ICS Press, San Francisco, USA

Ostrom E (1996) Crossing the great divide: Coproduction, synergy, development. World Dev 24(6):1073–1087

Ostrom E, Gardner R, Walker J (1994) Rules, games and common pool resources. Michigan State University, Ann Arbor, USA

Penczak T (1999) Fish population and food consumption in the Warta River (Poland): Continued post-impoundment study (1990–1994). Hydrobiologia 416:107–123

Penczak T, Glowacki L, Galicka W, Koszalinski H (1998) A long-term study (1985–1995) of fish populations in the impounded Warta River, Poland. Hydrobiologia 368(1–3):157–173

Prud'homme R (1994) On the dangers of decentralization. World Bank, Washington, DC, USA (Policy Research Working Paper 1252)

Przybylski M (1993) Longitudinal pattern in fish assemblages in the upper Warta River, Poland. Arch Hydrobiol 126(4):499–512

Przybylski M (1996) Variation in fish growth characteristics along a river course. Hydrobiologia 325(1):39–46

Ramu K (1999) River basin management corporation: An Indonesian approach. Paper presented at the Third River Basin Management Workshop, 23 June 1999, World Bank, Washington, DC, USA

Ramu K (2004) Brantas River basin, Indonesia. World Bank, Washington, DC, USA (Case Study Background Paper, available at www.worldbank.org/riverbasinmanagement)

Reynolds-Vargas JS, Richter DD Jr (1995) Nitrate in groundwaters of the Central Valley, Costa Rica. Environ Int 21(1):71–79

Ribot JC (2002) Democratic decentralization of natural resources: Institutionalizing popular participation. World Resources Institute, Washington, DC, USA

Rocha GA (2002) Um Copo d'Água. Editora Unisinos, Porto Alegre, Brazil

Rocha GA (2004) O Papel do Comitê de Bacia na Gestão das Águas. Presentation at the International Workshop on Geosciences for Urban Development, São Paulo, Brazil, August 2004

Rolla G (1998) Autonomy: A guiding criterion for decentralizing public administration. International Review of Administrative Sciences 64(1):27–39

SALASAN Associates Inc. (2002) Summary report: Evaluation of Fraser Basin Council effectiveness. Fraser Basin Council, Vancouver, British Columbia, Canada

Saleth RM, Dinar A (2004) The institutional economics of water: A cross-country analysis of institutions and performance. Edward Elgar, Cheltenham, UK

Samad M (2005) Water institutional reforms in Sri Lanka. Water Policy 7:125–140

Shaw PD, Tuominen T (1999) Water quality in the Fraser River basin. In: Gray C, Tuominen T (eds) Health of the Fraser River aquatic ecosystem. Aquatic and Atmospheric Sciences Division, Environment Canada, Vancouver, Canada (Department of Environment Fraser River Action Plan Report 1998-11)

Simon B (2002) Devolution of bureau of reclamation constructed facilities. J Am Water Resour As 38(5):1187–1194.

Teixeira FJC (2004) Modelos de Gerenciamento de Recursos Hídricos: Análises e Proposta de Aperfeiçoamento do Sistema do Ceará. Banco Mundial e Ministério da Integração Nacional, Brasília, Brazil (Série Água Brasil 6)

Tonderski A (1997) Control of nutrient fluxes in large river basins. Linköping University, Linköping, Sweden (Linköping Studies in Arts and Sciences 157)

Tonderski A, Blomquist W (2003) Warta basin. World Bank, Washington, DC, USA (Case Study Background Paper)

Usman R (2000) Integrated water resource management: Lessons from Brantas River basin in Indonesia. In: Abernathy C (ed) Intersectoral management of river basins. International Water Management Institute and German Foundation for International Development, Colombo, Sri Lanka (Proceedings of an International Workshop on Integrated Water Management in Water-Stressed River Basins in Developing Countries: Strategies for Poverty Alleviation and Agricultural Growth, Loksop Dam, South Africa, 16–21 October 2000)

Vermillion DL, Garces-Restrepo C (1998) Impacts of Colombia's current irrigation management transfer program. International Irrigation Management Institute, Colombo, Sri Lanka (Research Report)

World Bank (2003) Decentralizing Indonesia: A regional public expenditure review overview report. East Asia Poverty Reduction and Economic Management Unit, World Bank, Washington, DC, USA (Report No. 26191-IND)

World Bank (2005) India's water economy: Bracing for a turbulent future. World Bank, Washington, DC, USA (Sector Report No. 34750)

Wunsch JS (1991) Institutional analysis and decentralization: Developing an analytical framework for effective Third World administrative reform. Public Admin Develop 11:431–451

Additional Reading

Agarwal A, Narain S (1999) Making water management everybody's business: Water harvesting and rural development in India. International Institute of Environment and Development, London, UK (Gatekeeper Series No 87)

Ahmad N (1992) Decentralization: A study into forms, structures and development processes. Pakistan Academy for Rural Development, Peshawar, Pakistan

Bardhan P (1996) Decentralised development. Indian Economic Review 31(2):139–156

Bardhan P, Mookherjee D (1998) Expenditure decentralization and the delivery of public services in developing countries. Department of Economics, University of California, Berkeley, USA (Working Paper)

Bardhan P, Mookherjee D (1999) Relative capture of local and national governments. Department of Economics, University of California, Berkeley, USA (Working Paper)

Bardhan P, Mookherjee D (2000a) Capture and governance at local and national levels. AEA Papers and Proceedings 90:135–139

Bardhan P, Mookherjee D (2000b) Corruption and decentralization of infrastructure delivery in developing countries. Institute for Economic Development, Boston University, Massachusetts, USA (processed)

Bardhan P, Mookherjee D (2001) Decentralizing anti-poverty program delivery in developing countries. Institute for Economic Development, Boston University, Massachusetts, USA (processed)

Barrow CJ (1998) River basin development planning and management: A critical review. World Dev 26(1):171–186

Besley T, Coate S (1999) Centralized versus decentralized provision of local public goods: A political economy analysis. National Bureau of Economic Research, Cambridge, Massachusetts, USA (NBER Working Paper Series No 7084, available at *www.nber.org/papers/w7084*)

Bhat A, Ramu K, Kemper KE (2005) Institutional and policy analysis of river basin management in the Brantas River, East Java, Indonesia. World Bank, Washington, DC, USA (Policy Research Working Paper)

Bird RM (1994) Decentralization infrastructure: For good or for ill? World Bank, Washington, DC, USA (Policy Research Working Paper 1258)

Blomquist W, Ballestero M, Bhat A, Kemper KE (2004) Institutional and policy analysis of river basin management in the Tárcoles River, Costa Rica. World Bank, Washington, DC, USA (Policy Research Working Paper)

Blomquist W, Calbick KS, Dinar A (2004) Institutional and policy analysis of river basin management in the Fraser River, Canada. World Bank, Washington, DC, USA (Policy Research Working Paper)

Blomquist W, Giansante C, Bhat A, Kemper KE (2004) Institutional and policy analysis of river basin management in the Guadalquivir River, Spain. World Bank, Washington, DC, USA (Policy Research Working Paper)

Blomquist W, Haisman B, Dinar A (2004) Institutional and policy analysis of river basin management in the Murray-Darling River, Australia. World Bank, Washington, DC, USA (Policy Research Working Paper)

Blomquist W, Tonderski A, Dinar A (2004) Institutional and policy analysis of river basin management in the Warta River, Poland. World Bank, Washington, DC, USA (Policy Research Working Paper)

Chattopadhyay R, Duflo E (2001) Women as policy makers: Evidence from an India-wide randomized policy experiment. Massachusetts Institute of Technology, USA (unpublished paper)

Conyers D (1984) Decentralization and development: A review of the literature. Public Admin Develop (UK) 4:187–197

Cowie GM, O'Toole LJ (1998) Linking stakeholder participation and environmental decision making: Assessing decision quality for interstate river basin management. In: Coenen FHJM, Huitema D, O'Toole LJ (eds) Participation and the quality of environmental decision making. Kluwer Academic Publishers, Dordrecht, the Netherlands

Derman B, Ferguson A, Gonese F (2000) Decentralization, devolution and development: Reflections on the water reform process in Zimbabwe. Centre for Applied Social Sciences, University of Zimbabwe, Zimbabwe

Devas N (1997) Indonesia: What do we mean by decentralization? Public Admin Develop (UK) 17:351-367

Dourojeanni A (1994) Water management and river basins in Latin America. Cepal Review 53:111-128

Elamon J, Ekbal B (2000) Health sector reforms and local level planning: Experience of Kerala. International Conference on Democratic Decentralization, 23-27 May 2000, State Planning Board, Trivandrum, Kerala, India

Estache A, Sinha S (1994) Does decentralization increase public infrastructure expenditure? World Bank, Washington, DC, USA (Policy Research Working Paper 1457)

Evans P (1996) Government action, social capital and development: Reviewing the evidence on synergy. World Dev 24(6):1119-1132

Formiga Johnsson RM, Kemper KE (2005) Institutional and policy analysis of river basin management in the Alto Tietê River, Brazil. World Bank, Washington, DC, USA (Policy Research Working Paper)

Ghai D, Vivian JM (1992) Grassroots environmental action: People's participation in sustainable development. Routledge, London and New York

Ioris AAR (2001) Water resources development in the São Francisco River basin (Brazil): Conflicts and management perspectives. Water Int 26(1):24-39

IPECE (Instituto de Pesquisa do Estado do Ceará) (2003) Ceará em Números (available at *www.ceara.gov.br*)

Isaac TTM (2000) Campaign for democratic decentralisation in Kerala: An assessment from the perspective of empowered deliberative democracy. Kerala State Planning Board, Thiruvananthapuram, India (International Conference on Democratic Decentralisation, 23-27 May 2000)

Kemper KE, Formiga Johnsson RM (2005) Institutional and policy analysis of river basin management in the Jaguaribe River, Brazil. World Bank, Washington, DC, USA (Policy Research Working Paper)

Koppel B (1987) Does integrated area development work? Insights from the Bicol River basin development program. World Dev 15(2):205-220

Lam WF (1996) Institutional design of public agencies and coproduction: A study of irrigation associations in Taiwan. World Dev 24(6):1039-1054

Litvack J, Seddon J (1999) Decentralization briefing notes. World Bank Institute, World Bank, Washington, DC, USA (Working Paper)

Livingstone I, Assuncao LM (1993) Engineering vs. economics in water development: Dam construction and drought in Northeast Brazil. J Agr Econ 44:82-98

Mathew G, Nayak R (1996) Panchayats at work: What it means for the oppressed. Econ Polit Weekly July:1765-1771

McGinn N, Welsh T (1999) Decentralization of education: Why, when, what and how? International Institute for Educational Planning, United Nations Educational, Scientific and Cultural Organization, Paris, France (Fundamentals of Educational Planning Series No 64)

Merrey DJ (2000) Creating institutional arrangements for managing water-scarce river basins: Emerging research results. Paper prepared for presentation at session on Enough Water for All at the Global Dialogue on the Role of the Village in the 21st Century: Crops, Jobs, Livelihood, 15-17 August 2000, EXPO 2000, Hannover, Germany

Mestre ERJ (1997) Integrated approach to river basin management: Lerma-Chapala case study – attributions and experiences in water management in Mexico. Water Int 22(3):140-152

Mills A, Vaughan JP, Smith DL, Tabibzadeh I (1990) Health system decentralization: Concepts, issues and country experience. World Health Organization, Geneva, Switzerland

Mitchell B (1990) Integrated water management: International experiences and perspectives. Belhaven Press, London New York

Narayan D (1995) The contribution of people's participation: Evidence from 121 rural water supply projects. World Bank, Washington, DC, USA (Environmentally Sustainable Development Occasional Paper Series No 1)

Paterson J (1986) Coordination in government: Decomposition and bounded rationality as a framework for "user friendly" statute law. Aust J Publ Admin 45:95-111

Reynoso Vincente Guerrero LAE (2000) Towards a new water management practice: Experiences and proposals from Guanajuato State for a participatory and decentralized water management structure in Mexico. Water Resources Development 16(4):571-588

Romano PA, Cadavid Garcia EA (1999) Policies for water resources planning and management of the São Francisco River basin. In: Biswas AK, Cordeiro NV, Braga BPF, Tortajada C (eds) Management of Latin American river basins: Amazon, Plata and São Francisco. United Nations University Press, New York, USA

Salau AT (1990) Integrated water management: The Nigerian experience. In: Mitchell B (ed) Integrated water management: International experiences and perspectives. Belhaven Press, London New York

Schur M (2002) Pricing of irrigation water in South Africa (available at *www.worldbank.org/agadirconference*)

Scudder T (1989) The African experience with river basin development. Natural Resources Forum May:139–148

Shah A, Thompson T (2004) Implementing decentralized local government: A treacherous road with potholes, detours and road closures. In: Alm J, Martinez-Vazquez J, Mulyani Indrawati S (eds) Reforming intergovernmental fiscal relations and the rebuilding of Indonesia - The 'Big Bang' program and its economic consequences. Edgar Elgar Publishing

Sharma R (2003) Kerala's decentralisation: Idea in practice. Econ Polit Weekly XXXVIII(36): 3832–3850

Simpson LD (1999) The Rio São Francisco: Lifeline of the northeast. In: Biswas AK, Cordeiro NV, Braga BPF, Tortajada C (eds) Management of Latin American river basins: Amazon, Plata and São Francisco. United Nations University Press, New York, USA

Solanes M (1998) Integrated water management from the perspective of the Dublin Principles. Cepal Review 64:165–184

Stoffberg FA, van Zyl FC, Middleton BJ (1994) The role of integrated catchment studies in the management of water resources in South Africa. In: Kirby C, White WR (eds) Integrated river basin development. John Wiley and Sons, Chichester, UK

Tharakan PKM (2000) Community participation in school education: Experiments and experiences under people's planning campaign in Kerala. International Conference on Democratic Decentralization, 23–27 May 2000, State Planning Board, Trivandrum, Kerala, India

Tommasi M, Weinschelbaum F (1999) A principal-agent building block for the study of decentralization and integration. Universidad de San Andrés, Buenos Aires, Argentina (processed)

Udofia WE (1988) The role of river basins and rural development authorities in the development process. Third World Plan Rev 10(4):405–416

Usman R (2003) Comprehensive development of the Brantas River basin, the Republic of Indonesia. Paper presented at the 3rd World Water Forum, 16–23 March 2003, in Kyoto, Shiga, and Osaka, Japan

Visscher JT, Bury P, Gould T, Moriarty P (1999) International water resource management in water and sanitation projects: Lessons from projects in Africa, Asia and South America. IRC International Water and Sanitation Centre, Delft, the Netherlands

Walmsley N, Hasnip NJ (1997a) Case studies for water resource planning: Murray-Darling basin (Australia). Department for International Development, Wallingford, UK (Report OD/TN 89)

Walmsley N, Hasnip NJ (1997b) Case studies on water resource planning: Lessons learned and keys to success. Department for International Development, Wallingford, UK (Report OD 138)

Wandschneider PR (1984) Managing river systems: Centralization versus decentralization. Nat Resour J 24:1043–1066

White G (1998) Reflections on the 50-year international search for integrated water management. Water Policy 1(1):21–27

World Bank (1993a) China Yellow River basin investment planning study. World Bank, Washington, DC, USA

World Bank (1993b) Water resources management. World Bank, Washington, DC, USA (Policy Paper)

World Bank (1994) Infrastructure for development: World development report. World Bank, Washington, DC, USA

Zusman E (1998) A river without water: Examining water shortages in the Yellow River basin. LBJ Journal of Public Affairs (available at *uts.cc.utexas.edu/~journal/1998/river.html*)

Index

Printing: Krips bv, Meppel
Binding: Stürtz, Würzburg